IAEA NUCLEAR ENERGY SERIES No. NG

INTERNATIONAL NUCLEAR MANAGEMENT ACADEMY MASTER'S PROGRAMMES IN NUCLEAR TECHNOLOGY MANAGEMENT

INTERNATIONAL ATOMIC ENERGY AGENCY
VIENNA, 2020

COPYRIGHT NOTICE

© IAEA, 2020

Printed by the IAEA in Austria
August 2020
STI/PUB/1795

IAEA Library Cataloguing in Publication Data

Names: International Atomic Energy Agency.
Title: International Nuclear Management Academy Master's programmes in nuclear technology management / International Atomic Energy Agency.
Description: Vienna : International Atomic Energy Agency, 2020. | Series: IAEA nuclear energy series, ISSN 1995–7807 ; no. NG-T-6.12 | Includes bibliographical references.
Identifiers: IAEAL 20-01346 | ISBN 978–92–0–107217–7 (paperback : alk. paper) ISBN 978–92–0–116520–6 (pdf) | ISBN 978–92–0–116620–3 (epub) | ISBN 978–92–0–116720–0 (mobipocket)
Subjects: LCSH: Nuclear industry — Employees. | Nuclear engineering. | Degrees, Academic.
Classification: UDC 378:621.039 | STI/PUB/1795

FOREWORD

One of the IAEA's statutory objectives is to "seek to accelerate and enlarge the contribution of atomic energy to peace, health and prosperity throughout the world." One way this objective is achieved is through the publication of a range of technical series. Two of these are the IAEA Nuclear Energy Series and the IAEA Safety Standards Series.

According to Article III.A.6 of the IAEA Statute, the safety standards establish "standards of safety for protection of health and minimization of danger to life and property". The safety standards include the Safety Fundamentals, Safety Requirements and Safety Guides. These standards are written primarily in a regulatory style and are binding on the IAEA for its own programmes. The principal users are the regulatory bodies in Member States and other national authorities.

The IAEA Nuclear Energy Series comprises reports designed to encourage and assist research and development on, and application of, nuclear energy for peaceful uses. This includes practical examples to be used by owners and operators of utilities in Member States, implementing organizations, academia, and government officials, among others. This information is presented in guides, reports on technology status and advances, and best practices for peaceful uses of nuclear energy based on inputs from international experts. The IAEA Nuclear Energy Series complements the IAEA Safety Standards Series.

In 2011 and 2014, IAEA Nuclear Energy Series Nos NG-T-6.1, Status and Trends in Nuclear Education, and NG-T-6.4, Nuclear Engineering Education: A Competence Based Approach to Curricula Development were issued. These publications provide guidance to support the development of policies and strategies in nuclear education and to assist decision makers in Member States on a competence based approach for the development of curricula in nuclear engineering.

This publication follows the recommendations of NG-T-6.1 and NG-T-6.4 and presents the IAEA's International Nuclear Management Academy for universities that provide master's degree programmes that focus on management for the nuclear and radiological sectors. The overriding objective of the International Nuclear Management Academy is to improve the safety, performance and economics of nuclear and radiological technologies in Member States by promoting high quality master's level educational programmes for nuclear and radiological sector managers.

This publication is applicable to practitioners, decision makers and stakeholders of nuclear educational programmes from governments, academia, regulators, facility owners, operators and private industry. It introduces and describes nuclear technology management master's degree programmes. Nuclear technology management is relevant for newcomer countries that are embarking on nuclear energy programmes, countries with mature nuclear power development and countries with solely radiological facilities.

The IAEA would like to acknowledge the extrabudgetary support for the start-up of the International Nuclear Management Academy from the Ministry of Economy, Trade and Industry of Japan.

The IAEA wishes to acknowledge the valuable assistance provided by the contributors and reviewers listed in the contributors to drafting and review. The IAEA officers responsible for this publication were J. de Grosbois, F. Adachi and J.W. Roberts of the Division of Planning, Information and Knowledge Management.

CONTENTS

1. INTRODUCTION

1.1. BACKGROUND

The important role that the IAEA plays in assisting Member States in the preservation and enhancement of nuclear knowledge and in facilitating international collaboration in this area has been recognized by the IAEA General Conference in several resolutions, including Ref. [1]. These resolutions consider nuclear education and training as a prerequisite for the safe and efficient operation of nuclear and radiological facilities, and request the IAEA to assist Member States in their efforts to ensure the availability and sustainability of high quality nuclear education and training in all areas of nuclear technology for peaceful purposes, by focusing on the following:

— Developing policies and strategies in nuclear knowledge management, including key issues of nuclear education to help address national and regional needs;
— Fostering strong regional or interregional nuclear education networks;
— Facilitating the harmonization of curricula in nuclear education and training programmes;
— Promoting the awareness and use of nuclear and radiological facilities and engineering and training simulators as effective tools to enhance education and research and to maintain capability;
— Providing specific consultancy services through missions to address emergent problems, trends and long term issues relating to nuclear education;
— Supporting the development and adoption of innovative new e-learning technologies and pedagogical approaches to education and training;
— Creating awareness and common approaches to competency modelling and management frameworks;
— Analysing and sharing information and resources to facilitate nuclear education development.

The development of any national nuclear or radiological programme is dependent on the successful development of qualified human resources, through sustainable nuclear education and training programmes supported by government and industry. Among the broad range of specialists needed to ensure a capable nuclear and/or radiological workforce for the continued safe and economic utilization of nuclear technologies for peaceful purposes, competent nuclear managers are a vital component. Management of nuclear technology over the applicable life cycle is challenging, complex and requires managerial competencies specific to each nuclear or radiological sector. The safe and economic use of nuclear and radiological technologies relies heavily on many specialized disciplines and requires the systematic management of complex social and technical systems in a highly regulated environment.

Current and future nuclear and/or radiological sector managers in countries with, or planning to have, nuclear energy programmes or solely radiological programmes are expected to obtain the appropriate knowledge and skills for their positions, both nuclear technology related and managerial competencies. Ideally, future managers in the nuclear and radiological sectors should acquire most of the necessary competencies prior to moving into managerial positions and should continue their development while working as managers.

In many countries considering, or in the process of, launching nuclear energy programmes, or those utilizing radiological applications, there is a lack of both technical and managerial experience in management and leadership roles. Continuous professional development and on-the-job training programmes are in place in many countries, but the timelines needed to develop management leadership competency are longer than desired. In-house training may not be possible as it may be very costly and may not be as comprehensive as desired.

Engineers and scientists at nuclear and radiological organizations often have limited opportunities to obtain formal management education. Likewise, many managers in the nuclear and radiological sectors often do not have a qualification in a nuclear related technical degree programme and typically have few chances of obtaining such formal nuclear engineering or science education during their career. There therefore exists a need for managers to acquire management competencies specific to the nuclear and radiological sectors, not only through practical industry focused training courses and on-the-job learning, but also through formal education focused on theory, concepts and academic exercises.

Following interviews with managers working in the nuclear industry, and consultation with universities, it became clear that there is a need for master's degree programmes specializing in management for the nuclear and radiological sectors. Conventional university master's programmes in business administration, technology management and public administration provide extensive courses on various management aspects, but they do not explicitly address the specific and technology oriented managerial competencies necessary for the nuclear and radiological sectors. Nuclear regulators and licensed nuclear and radiological organizations also recognize the need, interest and benefits of establishing formal master's level educational programmes.

The IAEA has been working with nuclear sector organizations and universities to identify the competencies and expertise required by managers working in the nuclear industry. To develop a master's programme that is both flexible enough to provide the training required internationally and to accommodate a broad range of entry level experience, the IAEA consulted with managers to get feedback on education requirements and with a number of universities to design a suitable programme.

This collaboration led to the formation of the International Nuclear Management Academy (INMA) for universities that provide master's degree programmes focusing on management for the nuclear and radiological sectors. Through INMA, the IAEA facilitates collaboration with universities and other educational institutions in Member States that are authorized by their governments to confer master's degrees. It is supported as a programme activity of the Nuclear Knowledge Management Section of the IAEA's Department of Nuclear Energy. The IAEA facilitated INMA activities are regulated by the INMA Terms of Reference, listed in Appendix VIII.

The master's degree programmes must be high quality, consistent, tailored to address the requirements of the nuclear and radiological sectors, and, preferably, be available part time and by distance learning or short format courses to be accessible to currently employed nuclear professionals. Availability of the courses in English is encouraged, which will support internationalization of the nuclear workforce and help to meet the needs of developing countries.

INMA master's degree programmes in nuclear technology management (INMA-NTM programmes) are designed to meet these requirements and to provide managers with a broad understanding of nuclear technology and management best practices in a nuclear or radiological context. Specific managerial knowledge such as nuclear safety, security and safeguards; a global perspective on the nuclear and radiological issues; engineering economics; public relations and ethical issues aim to prepare the programme graduates to deal with risk informed decision making. The INMA-NTM programme curriculum specifically addresses the IAEA safety standards, nuclear security series, and guidance on safeguards on most topics. INMA-NTM programmes will facilitate an ongoing supply of highly qualified managers needed by nuclear energy sector employers, including nuclear power plants, waste management facilities, research and development laboratories, radiological facilities, regulatory bodies, technical support organizations, new build projects and nuclear energy related government ministries.

1.2. OBJECTIVE

The main objective of this publication is to provide an overview of the INMA master's degree programme for universities in IAEA Member States (i.e. government accredited) that provide master's degree programmes in technology management, nuclear science and engineering, general management including business management and public administration, plus those that already offer, or are interested in offering, master's level degree programmes in nuclear technology management.

In addition to detailing the common required elements of INMA-NTM degree programmes, this publication also outlines the assistance that is available for universities wishing to establish a master's degree programme in NTM, including the formal process of INMA-NTM programme endorsement. This publication is also for use by major nuclear and radiological sector employers and stakeholders, such as nuclear regulators, nuclear power utilities, technical support organizations, and nuclear research and development institutes that are expected to send or sponsor current or future managers (as registered students) to INMA-NTM programmes and who are expected to collaborate with INMA members by providing lecturers with industry experience, and feedback on industry needs or fellowship support. Some stakeholders who may already provide NTM related training can incorporate the INMA curriculum topics defined in this publication and may wish to collaborate with universities providing INMA-NTM programmes.

This publication is part of a series of IAEA publications dealing with nuclear education and the role of universities in building, strengthening, preserving and managing nuclear knowledge [2, 3]. For example, the IAEA publication Status and Trends in Nuclear Education [2] supports the development of policies and strategies in nuclear education and provides a review of the status of nuclear education in over thirty Member States and educational networks.

1.3. SCOPE

The following main issues are addressed in this publication:

— An overview description of the INMA initiative;
— The needs, interest and benefits of establishing formal master's level educational programmes, focusing on management for the nuclear sector including radiological applications;
— An assessment process to determine whether a university's NTM programme is INMA compliant;
— The endorsement process and INMA membership;
— Recommendations for implementation of INMA-NTM programmes at Member State universities and other educational institutions;
— Cooperation and collaboration of universities with industry and government to implement INMA-NTM programmes;
— National and international cooperation and educational networks;
— The role of the IAEA and activities that support nuclear education.

Guidance provided here, describing good practices, represents expert opinion but does not constitute recommendations made on the basis of a consensus of Member States.

1.4. STRUCTURE

Section 1 provides the background and objectives of this publication. It explains how competent managers are supported through formal educational programmes in NTM and why they are a vital component to sustaining the continued safe, economic utilization of all nuclear technologies for peaceful purposes. Section 2 presents a general description and the objectives of INMA, and gives a description of INMA-NTM programme implementation. Section 3 provides guidance on the development of INMA-NTM programmes based on fifty curriculum topics. It also explains how to tailor and design an individual INMA-NTM programme that provides sufficient flexibility to address the needs of local stakeholders and student needs, as well as accommodating possible programmatic themes. Section 4 presents the endorsement process for INMA-NTM with their associated missions, self-assessment tools and information packages. Sections 5, 6 and 7 describe the INMA web site, INMA annual meetings and the publication's conclusion, respectively.

Appendix I describes the suggested university course subjects for each INMA curriculum topic defined in Section 3. The description for each curriculum topic has been agreed by the INMA members. Appendix II shows the self-assessment tool and templates of the forms for use on an assist mission. Appendix III shows the self-assessment tool and templates of the forms for use on an assessment mission. Appendix IV is the template for the agenda of a mission. Appendix V contains suggestions to guide the preparation of an INMA mission. Appendix VI is the template for the assist mission report. Appendix VII is the template for the assessment mission report. Appendix VIII is the INMA Terms of Reference.

2. THE INTERNATIONAL NUCLEAR MANAGEMENT ACADEMY

2.1. GENERAL DESCRIPTION OF THE INTERNATIONAL NUCLEAR MANAGEMENT ACADEMY

Through the INMA initiative, the IAEA supports collaboration among nuclear engineering and science faculties and departments at universities in IAEA Member States to develop and implement INMA-NTM programmes. Any university or educational organization authorized by its government to confer a master's degree can establish an INMA endorsed master's degree programme in nuclear technology management. To have it endorsed as an INMA-NTM programme the university or educational organization must have their programme successfully assessed through an INMA mission (described in detail in Section 4.3).

The IAEA coordinates the following INMA activities in cooperation with INMA members, prospective members and stakeholders:

— Developing and reviewing the INMA-NTM programme curriculum topics;
— Organizing missions to support universities in evaluating the feasibility of implementing an INMA-NTM programme and building stakeholder support;
— Conducting missions to assess a proposed or existing university INMA-NTM programme;
— Coordinating an annual meeting for the INMA members and candidates for INMA membership.

Universities or educational organizations implementing INMA-NTM programmes with the supporting tools (described in detail in Section 4) offered by the IAEA, are expected to complete the following:

— Deliver NTM programmes that are INMA compliant;
— Share their experience in implementing INMA-NTM programmes and provide feedback on the further development of INMA;
— Share basic data of their INMA-NTM programme periodically (e.g. number of applications, new students, graduates; provide a breakdown by country, organization, years of experience, full or part-time study, gender and age demographics, drop-out rate and post-graduation employment status).

There may be a relatively low number of potential students enrolling in some INMA-NTM programmes as compared to other nuclear master's degree programmes. It may therefore be challenging for some INMA-NTM programmes to ensure that there is an adequate number of students each year to sustain the programme. Collaboration and resource sharing among INMA universities is therefore encouraged to minimize costs and better utilize resources. In addition, major employer stakeholders of an INMA-NTM programme, such as nuclear regulatory authorities, technical support organizations, nuclear power utilities, nuclear research and development laboratories, and institutes and reactor designers and vendors, are encouraged to send current or future managers on the programme to strengthen and improve their competencies. Employers can also collaborate with INMA members by providing lecturers with industry experience and example case studies, which provide opportunities for the students to obtain topical and practical knowledge. Valuable educational materials and lectures from stakeholders can also be shared among the INMA members.

Once INMA-NTM programmes have been implemented in several Member State universities, the goal will be to encourage the local recognition of NTM as a professional designation. The concept is to encourage appropriate national authorities (e.g. regulators or engineering societies) to award the designation of NTM professional to INMA-NTM programme graduates when they adequately meet additional criteria, including accumulating appropriate levels of work experience.

2.2. OBJECTIVE OF INMA

The objective of INMA is to improve the safety, performance and economics of the peaceful use of nuclear technology through the improvement of managerial competencies in the nuclear and radiological sectors. This objective is achieved by meeting many specific strategic goals relating to promoting and enabling the availability and accessibility of consistent, high quality and sustainable university master's level educational programmes for managers who work in nuclear or radiological organizations, or who are in nuclear policy and programme decision making roles in the government. The following are strategic aims to meet the objective of INMA:

— Supporting collaboration among nuclear engineering and science universities in IAEA Member States to develop guidance for implementing and delivering master's level educational programmes in NTM;
— Encouraging Member States to recognize the importance of nuclear management professionals in achieving and maintaining high levels of safety, security and performance;
— Ensuring that both the real needs of the nuclear and radiological sectors are met and that the IAEA safety standards, nuclear security series and guidance on safeguards are embedded as key teaching elements in the common requirements and the defined learning outcomes;
— Coordinating assessments of master's level programmes in NTM;
— Integrating nuclear industry experience with formal academic education;
— Encouraging various mechanisms, including university collaboration and resource sharing, e-learning, distance education, part-time programmes or short format courses, and innovative use of technology;
— Ensuring NTM programmes provide adequately balanced curricula that ensure coverage of the required curriculum topics in four categories: external environment, technology, management and leadership;
— Encouraging the local recognition of NTM professionals (as a professional designation), once INMA-NTM programmes have been implemented in one or more local Member State universities.

2.3. GENERAL DESCRIPTION OF INMA-NTM PROGRAMME IMPLEMENTATION

It is expected that for a university's NTM programme to be endorsed as an INMA-NTM programme it will be substantially (greater than 80%) based on the INMA curriculum topics, as assessed through an INMA assessment mission. The endorsement and university membership of INMA commences on receipt of a formal letter from the IAEA confirming the endorsement, accompanied by an INMA-NTM programme certificate.

Ensuring the quality of INMA-NTM master's programmes is critical for the students who enrol, the stakeholders who send their employees as students to the programmes, and the nuclear industry as a whole. The high quality of the INMA-NTM programmes will be maintained by the continual review of the INMA curriculum topics and the submission of the annual reports by the INMA members.

To provide assistance with the implementation of an INMA-NTM programme, a university can request an INMA assist mission. The purpose of an assist mission is to help the university (typically a nuclear department or faculty with an existing nuclear related course) to fully understand INMA-NTM programmes, to share the experiences of other INMA members, and to provide guidance on programme design. An assist mission also helps a university to determine the feasibility of implementing an INMA-NTM programme and to identify any further actions and resources that may be required.

When the university has completed the design of its INMA-NTM programme, a subsequent assessment mission to the university for possible endorsement by the IAEA can be organized. This will in many cases take place after the programme has been implemented, see Section 4 for more details. The assessment, which is presented in the form of a report prepared by the mission team, provides observations and recommendations for possible improvements based on the collective experience and interpretation of the team members.

In eligible countries, the IAEA Department of Technical Cooperation can offer programme support to develop and implement INMA-NTM programmes. The plans should be prepared well in advance with the awareness and support of the assigned programme management officer and appropriate national authorities, including the Member State national liaison officer. It is possible to embed these INMA-NTM programme development and support activities in the country's programme framework and national development plan.

3. INMA NUCLEAR TECHNOLOGY MANAGEMENT PROGRAMMES' CURRICULUM TOPICS

In this section, the curriculum topics for INMA-NTM programmes are specified and grouped into four categories. The INMA members, with facilitation by the IAEA, have provided descriptions of the curriculum topics. Further elaboration with suggested course subjects are provided in Appendix I.

An INMA master's programme builds knowledge and competence relating to nuclear technology management. A programmatic theme could be chosen by the university during the design of the programme to meet local stakeholder needs. Students are expected to exhibit some originality in the implementation of knowledge to show they are able to deal with complex issues both systematically and creatively. They are also expected to develop a knowledge of industry best practices and, in some areas, be at the forefront of a chosen professional discipline.

In general, the learning outcomes that most universities require to be awarded a master's degree include achieving an adequate depth and breadth of knowledge; demonstrating sufficient familiarity with the more common methodologies and some more advanced techniques in a possible area of specialization; demonstrating the ability to apply knowledge in the critical analysis of a new question or context, and to communicate ideas, issues and conclusions clearly; implementing an appropriate thesis or major project to demonstrate an acceptable level of research and scholarship; obtaining a reasonable awareness of the limits of knowledge in the area; and demonstrating some autonomy and professional capacity for management level analysis and decision making.

INMA-NTM master's degree programmes should install the character and leadership qualities, professionalism and transferable skills necessary for employment. This includes requiring the student to be capable of exercising initiative, personal responsibility and accountability (including for safety, security and safeguards), and the ethical behaviour consistent with academic integrity and a professional workplace.

3.1. CURRICULUM TOPICS

Fifty curriculum topics are specified that are expected to form the basis of any INMA-NTM programme. They are grouped into the four categories described below.

(1) External environment: The curriculum topics relating to understanding or managing aspects of the nuclear organization's external environment such as political, legal, regulatory, business and societal environments in which nuclear managers operate. Directly or indirectly, the external environment constrains, orients, influences or governs many decisions and actions of a nuclear manager.
(2) Technology: The curriculum topics relating to the basics of nuclear technology, engineering, and their applications that are involved directly or indirectly in the management of nuclear facilities for power and non-power applications.
(3) Management: The curriculum topics relating to the challenges and practices of management in the nuclear and radiological sectors with due consideration of safety, security and economics.
(4) Leadership: Requires an understanding of the technology and management of a nuclear facility with due consideration of the external environment in which it operates. Leadership requires vision, strong ethical behaviours, clear foresight and goal setting, commitment to safety and security, good communication skills with all stakeholders, and a professional disposition in all situations. Leaders in nuclear or radiological organizations are more effective when they have an understanding of high level technological competencies, coupled with strong managerial skills.

Category 1: External environment

The eleven curriculum topics in the external environment category are described below:

1.1. Energy production, distribution and markets: The global energy and nuclear energy environment in which nuclear organizations must remain competitive and efficient.

1.2.   International nuclear and radiological organizations: The purpose, roles, mission, goals, scope of operations, impacts, achievements, cooperation and interrelationships among key nuclear and/or radiological sector organizations, associations and stakeholder groups.

1.3.   National nuclear technology policy, planning and politics: The policy aspects that are required for the successful and ongoing operation of nuclear and/or radiological facilities. Due to the long lifetimes of nuclear facilities, any possible changes in energy policy and the political environment over time should be recognized, and how these factors may affect the viable operation of nuclear facilities should be understood.

1.4.   Nuclear standards: The existing international and domestic standards pertaining to the nuclear or radiological sector, including IAEA safety standards, other international standards and their influence on national regulatory requirements.

1.5.   Nuclear and radiological law: The framework of national legal obligations, international conventions, treaties, agreements and United Nations Security Council resolutions that establish the environment in which nuclear organizations and their managers must function.

1.6.   Business law and contract management: The legal context, including the liability and contractual issues and their management, taking into consideration lessons learned from previous legal interpretations and the enforceability of contracts. Resolution management is required for the successful implementation of some contracts.

1.7.   Intellectual property management: Intellectual property rights and their management, as well as associated issues, for nuclear facilities and projects across the nuclear and radiological sectors.

1.8.   Nuclear and radiological licensing, licensing basis and regulatory processes: The regulatory licensing processes for nuclear and/or radiological facilities and related processes and practices.

1.9.   Nuclear security: The national and international issues, frameworks, norms, obligations and approaches relating to physical nuclear security. The prevention and detection of, and response to, theft, sabotage, unauthorized access, illegal transfer or other malicious acts involving nuclear material, other radioactive substances as well as their practical implementation and impact on licensed nuclear or associated facilities.

1.10.   Nuclear safeguards: The national and international issues, frameworks, norms, obligations and approaches relating to nuclear safeguards as well as their practical implementation and impact on licensed nuclear facilities.

1.11.   Transport of nuclear goods and materials: The international and domestic frameworks for the safe transport of nuclear goods and materials, and the safety and security risk associated with their transport.

Category 2: Technology

The fifteen curriculum topics in the technology category are described below:

2.1.   Nuclear or radiological facility design principles: The technology, management, safety, security and non-proliferation influences on the principles of designing a nuclear power plant or other major nuclear or radiological facility.

2.2.   Nuclear or radiological facility operational systems: All major plant systems and operating modes of a typical nuclear or radiological facility as well as the main components of each facility or system, and their functions.

2.3.   Nuclear or radiological facility life management: The principles of long term operation and management of plant life. Nuclear facilities have special long term operational needs, given the level of their safety, reliability, economics and regulatory licence requirements.

2.4.   Nuclear or radiological facility maintenance processes and programmes: The strategies and methodologies that may be used to establish maintenance programmes at nuclear or radiological facilities, with various levels of complexity of licence conditions and design specifications. Planning and activity scheduling as a management tool is also required for an effective quality assurance programme.

2.5.   Systems engineering for nuclear or radiological facilities: The interdisciplinary approach enabling the successful design and implementation of complex systems. This includes defining the required functionality early in the development life cycle, documenting requirements and then proceeding with design synthesis and system validation, which delivers successful implementation (i.e. to achieve equipment system reliability, safety, performance, economics or other goals). An understanding is required of how system engineering plays an important role throughout the life cycle of the nuclear or radiological facility.

2.6.   Nuclear safety principles and analysis: The nuclear safety fundamentals, their principles, analysis methods and how the industry is responding to the safety requirements.

2.7.   Radiological safety and protection: The justification and optimization of protection for planned, emergency and existing exposure situations. Safety and protection are ensured by the understanding of the radiological impact on different materials and on the human body. Substantial control on each nuclear site is maintained by adherence to the radiological regulations in each country.

2.8.   Nuclear reactor physics and reactivity management: The general awareness of the physical principles needed for the design, construction and operation of nuclear reactors. This also includes all the factors that affect reactivity and criticality.

2.9.   Nuclear fuel cycle technologies: The entire fuel cycle, from mining to final disposal. It includes all the technologies used to produce and manage nuclear materials as well as the security of the fuel supply, reduction in fuel cycle costs, management of the waste streams, and non-proliferation issues.

2.10.   Radioactive waste management and disposal: The technologies associated with radioactive waste management including the stringent controls on radiological releases (solids, liquids, airborne materials, gases) to the environment. It ensures all measures are adhered to regarding the safety of people and the environment.

2.11.   Nuclear or radiological facility decommissioning: The technological challenges and regulatory aspects associated with the end of the operating life of a nuclear or radiological facility. Consideration should be given to the various decommissioning strategies that have alternative financial and radiological implications.

2.12.   Environmental protection, monitoring and remediation: The hydrological and ecological effect of nuclear and radiological facilities during normal operation through to decommissioning, and from the consequence of accidents on the local and wider environment. This curriculum topic also covers the effect of contamination of the local environment from radionuclides due to historical discharges, accidental releases, as well as releases from non-nuclear facilities such as mining, oil, gas and medical industries, and any subsequent environmental remediation.

2.13.   Nuclear research and development and innovation management: The technical innovation processes in nuclear and radiological sectors and industry research and development. This includes familiarization and knowledge of the role and scope of nuclear research and development organizations and their specific challenges. Knowledge of the current trends in next generation nuclear technologies provides an understanding of the current timeline for the deployment of various innovations.

2.14.   Applications of nuclear science: The range of applications of nuclear science in the fields of research, medicine, industry, food and agriculture, the environment, security and space exploration.

2.15. Thermohydraulics: The general knowledge of the physical principles needed to ensure adequate cooling of nuclear fuel during operation, shutdown, accident conditions and long term storage. This includes cooling of containment and consideration of ultimate heat sinks.

Category 3: Management

The eighteen curriculum topics in the management category are described below:

3.1. Nuclear engineering project management: The management of construction, refurbishment and decommissioning projects. This involves the various specifics of small, medium and large design and build projects and related procurement. Key aspects of any nuclear project include decision making processes, worker qualification, planning, project monitoring and control, documentation and communication management. Risk management, supplier quality control, licensing processes and major stakeholder roles and responsibilities should also be included.

3.2. Management systems in nuclear or radiological organizations: Various management systems in nuclear or radiological organizations include, for example, an operations management system, a training management system, a supplier management system, a quality management system, a work management system, an outage planning system and a licensing or regulatory compliance system. This curriculum topic emphasizes the importance of management systems to ensure that work processes are planned, monitored and controlled in a safe and systematic manner. It also acknowledges the importance of an integrated approach.

3.3. Management of employee relations in nuclear or radiological organizations: The aspects to ensure a collaborative and cooperative relationship with and between employees, including contractual (i.e. collective labour agreement) considerations, emotional and trust issues, physical and practical aspects. It includes consideration of how employee relations may be influenced by local employment laws and cultural norms. A key focus is how to create an appropriate work environment that supports the safety and economic objectives of the organization. Factors of organizational culture, accountability and workforce performance should be addressed (e.g. establishing trust through open communications and a supportive culture of knowledge sharing and promoting positive gender relations).

3.4. Organizational human resource management and development: The aspects of the employment cycle. This curriculum topic also includes performance evaluation systems, workforce planning and adjustments, and ensuring the workforce can adequately meet its human resource responsibilities. Other factors to be addressed include succession planning, salary and benefits administration and organizational structure and performance.

3.5. Organizational behaviour: The theory and concept of organizational behaviour in the context of the nuclear and radiological sectors and in particular its potential impact on safety, security and performance. This includes consideration of the interaction between individuals and work groups within the organizational structure and setting. This curriculum topic includes issues relating to interdependencies of organizational behaviour along with other aspects of management (e.g. influence of stakeholders, leadership, organizational culture).

3.6. Financial management and cost control in nuclear or radiological organizations: The financial aspects and related risks associated with nuclear operations or projects and the importance of cost control in the effective management of budgets, scheduling and resources.

3.7. Information and records management in nuclear or radiological organizations: The requirement to understand processes, applications, roles, responsibilities and challenges involved in information and records management in the nuclear and radiological sectors.

3.8. Training and human performance management in nuclear or radiological organizations: The training and human performance management aspects, including ensuring that individuals have the competencies needed to perform their assigned tasks, organizing work effectively, and monitoring and continually improving performance.

This includes knowledge about the basic principles and tools for excellence in human performance and how those tools should be effectively integrated into all ongoing processes and programmes at a facility to ensure the desired results. The consideration may include performance improvement models, human performance improvement frameworks, nuclear facility personnel training, manager obligations and responsibilities, and the systematic approach to training as a management tool. The impact of any individual's job performance should be considered in thinking about the relevance and importance of human performance for the facility's operation, safety and security. The importance of training on cultural awareness and cross cultural communication should be acknowledged.

3.9. Performance monitoring and organization improvement: The current performance of an organization, detecting any subtle decline in performance, and looking at all opportunities for improvement by means of self-assessment, performance monitoring, external assessment and independent oversight. It focuses on the ongoing assurance that the management systems at nuclear organizations are effective at ensuring that licensed facilities remain demonstrably within their licence conditions and operate in the safest, most reliable and most cost effective manner possible.

3.10. Nuclear quality assurance programmes: The principles and approaches of quality management systems (i.e. quality assurance, quality control programmes) and their requirements, adoption and implementation as an essential part of an effective management system. Consideration is required as to how they should be applied to all activities affecting the processes and services important to the safety, reliability, performance and security of the facility. This competence also addresses concepts and approaches to the successful implementation of quality assurance systems including planned and systematic actions that provide adequate confidence that the specified requirements are satisfied.

3.11. Procurement and supplier management in nuclear or radiological organizations: The management of the procurement process and the relationships with the suppliers to the nuclear organizations, which have a direct impact on quality assurance at the nuclear facility. All items and services procured must have specified quality requirements to ensure that they do not have an adverse impact on the safety or the operation of the nuclear facility. Due consideration will be required to comply with local procurement procedures.

3.12. Nuclear safety management and risk informed decision making: The ongoing consideration of the management of safety in the context of the management systems. Considerations include proper operating conditions, prevention of accidents, mitigation of accident consequences in order to protect workers, the public and the environment from radiation hazards. There are many sources, locations and hazards of radiation. Knowledge is required for the safe control of radiation hazards in nuclear installations, radioactive waste management and in the transport of radioactive material. This area also addresses management roles and responsibilities to ensure that effective decision and work processes are in place and that adequate organizational resources and accountabilities are established in a manner that specifically addresses safety concerns. These include management of risks under normal circumstances and the consequences of incidents and events that may lead to possible radiological release.

3.13. Nuclear incident management, emergency planning and response: The emergency preparedness and management of nuclear events and the associated emergency response in case of an accident. Integration of safety, security and emergency preparedness programmes provides the optimum protection for public health and safety.

3.14. Operating experience feedback and corrective action processes: The managed processes to identify the root causes of past events and prevent the recurrence of similar events. Operating experience is a valuable source of information for learning about and improving the safety, reliability and security of nuclear installations. It focuses on detecting and recording deviations from normal performance by systems and by personnel, especially those which could be precursors of events. It is essential to collect such information in a systematic way for events occurring at nuclear installations during commissioning, operation, maintenance and decommissioning. Corrective action processes are to develop the actions needed to prevent the recurrence of similar events and track their closure and their effectiveness with the aim of continuous improvement. Consideration should also be given to the role of international organizations such as the IAEA, the OECD Nuclear Energy Agency (OECD/NEA), the World

Association of Nuclear Operators (WANO) and Institute of Nuclear Power Operations that provide standards and guidelines in this regard.

3.15. Nuclear security programme management: The management practices in place to ensure that a nuclear site's security programme is implemented correctly through individual security responsibilities, regulatory compliance and event reporting. The risks to any particular site or their associated facilities will cover all perceived security threats from theft, sabotage, unauthorized access, illegal transfer or other malicious acts involving nuclear material, other radioactive substances, through to protestor disruption, cyberattacks and terrorism.

3.16. Nuclear safety culture: The influences of different factors such as organizational cultural values and norms, individually shared beliefs and perceptions and their impact on the organizational safety and performance. Methods to establish a positive organizational culture should be addressed and included in the training of new employees and the proactive fostering of safety awareness in all employees. A key component of the organizational culture will be the safety culture that requires continuous engagement and dialogue.

3.17. Nuclear events and lessons learned: The key international lessons learned from previous major historical nuclear accidents and their influence on current designs and operational procedures. Awareness and a cursory review of previous major accidents should reinforce the importance of nuclear safety, security and the understanding of hazards and consequences. Specific case studies are encouraged (e.g. the Fukushima Daiichi nuclear accident, the Chernobyl accident, the Three Mile Island accident, and other events).

3.18. Nuclear knowledge management: The establishment of a knowledge management programme and culture within an organization that aligns with a national capacity building policy and strategy, as well as an integrated part of the organizational nuclear infrastructure. This should include topics such as developing knowledge and skills to critically appraise the nature of nuclear knowledge and benefits for the safe operation of nuclear facilities, gains in economics and operational performance, facilitating innovations, and ensuring the responsible use of sensitive knowledge. Consideration should be given to the importance of treating knowledge as an asset, the potential benefits of applying knowledge management tools and techniques, approaches and practices to manage nuclear knowledge, specialized nuclear related information resources and appropriate knowledge management methods and tools.

Category 4: Leadership

The six curriculum topics in the leadership category [4–7] are described below:

4.1. Strategic leadership: The implementation of the overarching policies of a nuclear or radiological organization. These policies may be aligned with corporate or national strategies and managed by the leadership team, incorporating a strong nuclear safety and security culture. A leader should demonstrate the ability and discipline to solve problems, evaluate options, make judgements and implement a plan using skilful reasoning, accurate information, training and experience. A leader should demonstrate leadership for commitment to safety and security. Effective leadership in nuclear or radiological organizations requires a professional attitude and excellent interpersonal skills.

4.2. Ethics and values of a high standard: The principles and behaviours that underpin the decisions, strategies and values embodied in nuclear leadership. Leaders must establish and maintain a strong nuclear safety and security culture in developing good leadership at all levels for their organization. Leaders must foster an environment that promotes accountability and enhances safety performance and security.

4.3. Internal communication strategies for leaders in nuclear or radiological organizations: Internal communication should be consistent to ensure that the whole workforce understands their role in the attainment of organizational goals. Leaders must communicate clearly the basis for decisions relevant to safety, security and performance.

4.4. External communication strategies for leaders in nuclear or radiological organizations: External communication should be clear and transparent and developed with stakeholders, recognizing the prevailing environment affecting the operation of an organization (e.g. socioeconomic, political).

4.5. Leading change in nuclear or radiological organizations: A plan that is aligned to the strategic vision. Effective communication is required to ensure that the risks and opportunities relating to change are understood by all stakeholders, and how they may affect the safety and security of the nuclear organization. Examples may include internal reorganization, merger, acquisition or joint venture.

4.6. Leadership to support the safety culture: An environment that promotes accountability and enhances safety performance and security. A leader should demonstrate leadership for safety and a commitment to safety and security.

3.1.1. Suggested course subjects for the curriculum topics

Suggested course subjects for each curriculum topic (presented in Appendix I) have been agreed upon by the INMA members as a reference for universities. Universities can add more course subjects or omit some of them accordingly. Most of the suggested course subjects for the required curriculum topics are expected to be included in the universities' INMA-NTM programmes.

3.2. COMPOSITION OF INMA-NTM PROGRAMMES

INMA provides sufficient flexibility to enable each university programme to be tailored to the needs of local stakeholders and students, as well as to accommodate possible programmatic themes. Programme design decisions should be justified and based on these needs. Universities may identify a programmatic theme for their master's degree in NTM. This does not require universities to create and approve a new degree programme; they can work within their existing master's degree programmes. The breadth and depth of coverage of each of the curriculum topics are at the discretion of the university.

3.2.1. Breadth of coverage of the curriculum topics

It is expected that a university's INMA-NTM programme will include all of the 36 required curriculum topics (see Table 1), 72% of the total, and achieve 80–85% coverage of all the curriculum topics, primarily through coursework and project work. However, if an internship or practicum is incorporated as a formal part of the programme, this may be credited as contributing to achieving the expected overall learning outcomes.

INMA-NTM programme graduates are expected to have knowledge and critical awareness of current problems, and to have new insights. They are expected to be able to critically evaluate current and advanced research, consider complex issues based on established principles and techniques, and show originality in the application of knowledge. They are expected to be able to implement knowledge in the critical analysis of a new question or context, and communicate ideas, issues and conclusions clearly. They are also expected to demonstrate some autonomy and professional capacity for management level analysis and decision making.

3.2.2. Suggested learning hours

The typical envelope of total learning hours for INMA-NTM programmes at graduation is expected to be in the range of 1800–4000 hours. This wide range of learning hours acknowledges the different duration of a master's degree in different countries and allows for flexibility. It is based on consultations with university academic staff from different Member States. Learning hours are intended to encompass all contact hours (hours spent in contact in real-time with an instructor) and independent study hours.

The total learning hours are suggested to be achieved through a combination of lectures, self-study, research project or master's thesis and possibly an experiential component comprised of any combination of work based experience, on-the-job training and internships. Experiential learning during the programme may only be possible for part-time students or during the project phase for full time students, if it is based in industry.

TABLE 1. 'REQUIRED' OR 'AS APPROPRIATE' CURRICULUM TOPICS

	Curriculum topics[a]	
	1.1. Energy production, distribution and markets	A[b]
	1.2. International nuclear and radiological organizations	R[c]
	1.3. National nuclear technology policy, planning and politics	A
	1.4. Nuclear standards	R
	1.5. Nuclear and radiological law	A
Category 1: External environment	1.6. Business law and contract management	R
	1.7. Intellectual property management	A
	1.8. Nuclear and radiological licensing, licensing basis and regulatory processes	R
	1.9. Nuclear security	R
	1.10. Nuclear safeguards	A
	1.11. Transport of nuclear goods and materials	A
	2.1. Nuclear or radiological facility design principles	R
	2.2. Nuclear or radiological facility operational systems	R
	2.3. Nuclear or radiological facility life management	A
	2.4. Nuclear or radiological facility maintenance processes and programmes	R
	2.5. Systems engineering for nuclear or radiological facilities	A
	2.6. Nuclear safety principles and analysis	R
	2.7. Radiological safety and protection	R
Category 2: Technology	2.8. Nuclear reactor physics and reactivity management	A
	2.9. Nuclear fuel cycle technologies	A
	2.10. Radioactive waste management and disposal	R
	2.11. Nuclear or radiological facility decommissioning	R
	2.12. Environmental protection, monitoring and remediation	R
	2.13. Nuclear research and development and innovation management	A
	2.14. Applications of nuclear science	A
	2.15. Thermohydraulics	A

TABLE 1. 'REQUIRED' OR 'AS APPROPRIATE' CURRICULUM TOPICS (cont.)

	Curriculum topics[a]	
	3.1. Nuclear engineering project management	R
	3.2. Management systems in nuclear or radiological organizations	R
	3.3. Management of employee relations in nuclear or radiological organizations	R
	3.4. Organizational human resource management and development	R
	3.5. Organizational behaviour	R
	3.6. Financial management and cost control in nuclear or radiological organizations	R
	3.7. Information and records management in nuclear or radiological organizations	R
	3.8. Training and human performance management in nuclear or radiological organizations	R
	3.9. Performance monitoring and organization improvement	R
Category 3: Management	3.10. Nuclear quality assurance programmes	R
	3.11. Procurement and supplier management in nuclear or radiological organizations	R
	3.12. Nuclear safety management and risk informed decision making	R
	3.13. Nuclear incident management, emergency planning and response	R
	3.14. Operating experience feedback and corrective action processes	R
	3.15. Nuclear security programme management	A
	3.16. Nuclear safety culture	R
	3.17. Nuclear events and lessons learned	R
	3.18. Nuclear knowledge management	R
	4.1. Strategic leadership	R
	4.2. Ethics and values of a high standard	R
Category 4: Leadership	4.3. Internal communication strategies for leaders in nuclear or radiological organizations	R
	4.4. External communication strategies for leaders in nuclear or radiological organizations	R
	4.5. Leading change in nuclear or radiological organizations	R
	4.6. Leadership to support the safety culture	R

[a] The description of each curriculum topic is presented in Section 3.1 and elaborated with suggested course subjects in Appendix I.
[b] A — Indicates curriculum topic is included when appropriate depending on the theme or focus of a specific INMA-NTM programme.
[c] R — Indicates curriculum topic is required.

The recommended breakdown of the total learning hours for INMA-NTM programmes at graduation is shown in Table 2, but some deviations may be justified to meet stakeholder needs, or to consider incoming student cohort

TABLE 2. THE RECOMMENDED BREAKDOWN OF TOTAL LEARNING HOURS FOR NUCLEAR TECHNOLOGY MANAGEMENT PROGRAMMES AT GRADUATION

Type of learning	Approximate learning envelope (hours)
Courses: Combination of lectures and self-study	
Category 1: External environment	150–450
Category 2: Technology	450–750
Category 3: Management	300–750
Category 4: Leadership	150–300
Experiential: Any combination of work experience, on-the-job training and internships	600–1200
Research project or master's thesis preparation and evaluation	300–600

capability[1]. The total learning hours envelope should also be consistent with local accreditation requirements for master's programmes.

3.3. RECOMMENDED PROGRAMME ELEMENTS

3.3.1. Programme components

Based on a series of consultancy meetings from 2014 to 2016 to develop INMA, the following components have been identified as useful for INMA-NTM programme curricula:

(a) Group projects: These are strongly encouraged to be incorporated in INMA-NTM programmes. Master's level projects in which students are assigned real problems in the nuclear or radiological sectors are beneficial to foster all the elements of competencies, particularly the implementation level of knowledge. Students often gain insight from their fellow students during group projects and also build close networks with them.
(b) Lectures from industry: Close cooperation between universities and industry is an important factor in improving nuclear education and training. Lectures from professionals and managers with industry experience and the incorporation of real world example case studies can provide rich learning opportunities for students to obtain topical, specialized, context specific and practical knowledge.

3.3.2. Incorporating existing management schools and courses

Some institutes have developed excellent courses and programmes on NTM that could be included in INMA-NTM programmes. The programmes below are either organized by the IAEA, in cooperation with the IAEA, or by third parties for one to several weeks of duration. Universities are encouraged to consider incorporating these existing courses into their programmes. Universities will need to assess their students' requirements, how these courses match their programme content, and the benefits to be obtained from attendance.

---

[1] Some universities have processes in place to assess the experience and prior learning of incoming graduate level students and may grant a student an exemption or equivalent credit for one or more of their programme courses or modules. Some universities allow students (on an exceptional permission basis) to write only the final exam (in specific graduate courses) to obtain credits. These policies are at the university's discretion and may allow a student to better tailor their master's study programme to focus on areas of interest or the programmatic theme and achieve greater overall learning outcomes and benefits.

The following list includes examples of existing courses on NTM topics:

— IAEA Nuclear Energy Management School: Typically a 1–3 week programme that includes a range of nuclear management lectures, team projects and site visits to develop future leadership capability to manage nuclear energy programmes. The school targets young professionals from any organization relating to the nuclear sector. The IAEA together with the Abdus Salam International Centre for Theoretical Physics in Italy introduced the school in 2010. The school in Trieste is open to applicants from all Member States and is held annually. Regional schools are also held.

— IAEA Nuclear Knowledge Management School: Typically a 1–2 week programme that provides specialized education on knowledge management in nuclear science and technology organizations. It teaches methodologies and practices using tools, real life examples and good practices of different types of nuclear organizations. The IAEA introduced this school in 2004, together with the Abdus Salam International Centre for Theoretical Physics in Italy. The school in Trieste is open to applicants from all Member States. Regional schools are also held.

— World Nuclear University Summer Institute: An intensive, six-week leadership programme held annually since 2005 that offers a comprehensive programme of cutting edge presentations on a range of nuclear topics, a week of visits to different nuclear installations, dynamic team projects led by mentors, and networking opportunities with nuclear global leaders. It targets nuclear professionals from industry, regulatory bodies, educational and research organizations with demonstrated leadership potential and academic or professional excellence. The World Nuclear University Summer Institute is organized with support of the IAEA, the World Nuclear Association, WANO and OECD/NEA.

— Nuclear Law Institute: Established by the IAEA to meet the increasing demand for legislative assistance by Member States. Every year, the institute offers intensive training for two weeks in Vienna for up to 60 lawyers in all areas of nuclear law and in drafting corresponding national legislation. Due to the increasing number and complexity of international instruments adopted in the areas of nuclear safety, security, safeguards and liability, and to better meet the demand from Member States for legislative assistance, specifically for training and for capacity building, the IAEA Office of Legal Affairs decided in 2011 to streamline its legislative assistance activities by establishing a Nuclear Law Institute, in cooperation with the Department of Technical Cooperation.

— International School of Nuclear Law: Established in 2001 by the OECD/NEA, in cooperation with the University of Montpellier, the International School of Nuclear Law has been designed to provide participants with a comprehensive understanding of the various interrelated legal issues relating to the safe, efficient and secure use of nuclear energy. The programme has evolved over the last decade to address developments in nuclear law, thus providing a high quality, intensive overview of a complex body of laws and legal regimes.

— IAEA e-learning courses: The IAEA has developed many e-learning courses and many of them fit well with some of the detailed teaching elements for each curriculum topic. Some e-learning courses cover the IAEA's Milestones approach to introducing a nuclear power programme [8], which consists of modules on developing a human resource strategy, stakeholder involvement, management of a nuclear power programme, construction management and the systematic approach to training.

— Intercontinental Nuclear Institute programme: Developed in cooperation with the Czech Technical University in Prague and the University of Massachusetts, Lowell, and supported by the IAEA, the programme provides experiential learning supported through subject matter experts in reactor physics, design features, planning, licensing, operations, engineering, management, economics, safety and security, radiation detection and measurement, the nuclear fuel cycle, non-proliferation, and nuclear security and physical protection.

4. THE INMA ENDORSEMENT PROCESS AND MISSIONS

INMA missions help universities to understand, design and establish INMA-NTM programmes and to formally assess them with a view to endorsement, leading to the university becoming an INMA member. The preparation for the missions, including the pre-mission information packages, the format of the missions and the endorsement process, are described below.

4.1. OUTLINE OF THE ENDORSEMENT PROCESS

The activities to perform an assessment of a university's NTM programme for INMA-NTM programme compliance are listed below in chronological order.

Step 1: A university formally requests, in a formal letter through the Member State's official channels, an INMA mission to assist with the establishment of an INMA-NTM programme, through discussions with the organization's staff and stakeholders, and to identify curriculum topics not covered by its existing programme(s).

Step 2: Following acceptance of the request by the IAEA, the university, in accordance with the INMA Terms of Reference (see Appendix VIII), commences its preparations for the assist mission by conducting a self-assessment of its newly designed or existing programme(s) and submitting the preliminary information package to the IAEA.

Step 3: The INMA mission to assist the university is implemented, and the report, compiled on conclusion of the mission, is provided to the university.

Step 4: The university develops and establishes an NTM programme.

Step 5: The university requests an INMA mission to assess their programme by directly contacting the IAEA Nuclear Knowledge Management Section.

Step 6: The university performs a self-assessment of its NTM programme and submits a complete assessment information package consisting of a detailed programme description and background information for submission to the IAEA.

Step 7: The INMA mission to assess the NTM programme is implemented, and the assessment report is provided to the university.

Step 8: The university provides feedback on the main findings of the mission and recommendations made in the report, if any.

Step 9: The IAEA issues the assessment mission report. If no action plan is required, or only minor issues were found, the assessment report may recommend the university for INMA membership.

Step 10: The university responds to the IAEA with an action plan, if required, to address any major revisions or additions required of their programme and informs the IAEA once all issues have been addressed.

Step 11: The Director of the IAEA's Department of Nuclear Energy's Division of Planning, Information and Knowledge Management, in consultation with the Nuclear Knowledge Management Section Head, decides whether the IAEA accepts the recommendation for INMA membership and the endorsement of the university's NTM programme as an INMA-NTM programme.

Step 12: A formal letter signed by the IAEA, together with an INMA-NTM programme certificate, is sent to the university endorsing its NTM programme as an INMA-NTM programme. Until such letter is received, the university should not assume that it is an INMA member.

4.1.1. INMA-NTM programme self-assessments and information packages

Prior to any INMA mission, the university is requested to complete a self-assessment of its relevant programmes and complete the appropriate information package. Ideally, the information package should be prepared and submitted to the IAEA at least two months before the mission. This will enable the experts selected for the mission team, as well as the IAEA staff, to fully review the material before the mission and ask for any supplementary information if necessary. All information received and retained by team members will be subject to strict control by the IAEA and will not be released to others without the written consent of the university. The information package tools and forms are available for members to download from the INMA web site or can be requested from the IAEA Nuclear Knowledge Management Section.

4.1.2. Mission team's composition

The INMA mission teams usually include one or two IAEA staff members, plus experts who have experience in the development of an INMA-NTM programme at their university or have had significant input into a university's INMA-NTM's programme or the establishment of INMA. The team may also include observers from universities considering establishing an INMA-NTM programme. It is expected that all the experts on the mission team will be familiar with the INMA methodology and the review process. All the INMA members are encouraged to provide qualified experts to participate in missions.

4.1.3. The duration of the missions

The duration can be up to five working days depending on the scope of the mission. A longer mission may be appropriate if multiple locations or sites are to be visited, multiple degree programmes are being reviewed, or if broader national educational issues are to be discussed.

4.1.4. The responsibility of the local host

The local host is the main contact in the university hosting the mission. In general, the local host is responsible for the following activities:

— Making initial contact with the IAEA to arrange the mission. Informal preliminary discussions should be held with representatives within the IAEA's Nuclear Knowledge Management Section describing current issues and problems. The request for an assist mission must be addressed to the Deputy Director General, Department of Nuclear Energy.
— Specifying options for the date, time and place of the proposed mission meetings and presentations.
— Collating and sending to the IAEA the preliminary and assessment information packages prior to the appropriate mission.
— Coordinating travel and accommodation arrangements for all the mission team members and assisting others who may be attending meetings and presentations; ensuring security, health and safety, and welfare aspects.
— Arranging a hospitality event and technical tour (both are optional).
— Ensuring that equipment and meeting rooms are available to support presentations and meetings.
— Liaising with key internal and external stakeholders to ensure attendance at meetings and presentations, as appropriate.
— Ongoing communication with the IAEA on completion of the mission to provide any further information required by the mission team and provide feedback on the value of the mission.

4.2. INMA ASSIST MISSIONS

An INMA assist mission can be requested by a university to promote a better understanding of INMA-NTM programmes, support the self-assessment of their present nuclear management educational capacity and assist the development of their own INMA-NTM programme. An INMA assist mission can only be initiated after the IAEA receives a formal request from the university through the Member State's official channels. An assist mission can also support universities to develop a network between the university and the stakeholders.

Depending on the available resources and collaborations, the university may wish to develop their NTM programme with a specific theme, such as new build, decommissioning or applications of nuclear science. This will influence the aims for an assist mission, which are listed below:

— To provide a realistic time frame for establishing and implementing the university's INMA-NTM programme;
— To determine the equipment, facilities and human resource requirements for the programme's implementation;
— To provide guidance on the input available from industry that would strengthen the programme;
— To ascertain whether collaboration with other universities is initially required.

The assist mission team comprises INMA experts with one or two IAEA staff members. Additionally, some observers can be included if both the IAEA Scientific Secretary and the host university agree. The university that requests an assist mission will complete the preliminary information package (see Appendix II) described in Section 4.2.1 below, and return it to the IAEA in advance of the assist mission. It includes a self-assessment tool for the university's existing master's programmes that is used to map the coverage of existing courses and programmes and to identify gaps. This self-assessment helps to identify what would be needed to introduce an INMA-NTM programme even before the university receives the assist mission. The mission team will then be able to discuss and advise on the university's vision and approach.

The mission will consist of presentations by the host university's staff on courses that may form part of the proposed INMA-NTM programme, complemented by presentations on the pedagogical methodologies, resources that will be used to deliver the programme, visits to facilities and presentations from the INMA experts on their experiences implementing INMA-NTM programmes. It is also very beneficial to have contributions from stakeholders, particularly nuclear organizations that could be involved in the programme through provision of lectures or by sending their employees to study part time.

Collaboration between the university and stakeholders is important to focus the university's programme on local needs and to be effective. Help and support to develop stakeholder relationships can also be provided, if required.

4.2.1. Preliminary information package

The preliminary information package consists of the INMA preliminary self-assessment tool and the INMA preliminary programme description form, both shown in Appendix II.

4.2.1.1. INMA preliminary self-assessment tool

The INMA preliminary self-assessment tool is designed for universities to map their current relevant programmes that could provide courses or content for their NTM programme against the INMA curriculum topics. Universities are requested to assess their capability to deliver each curriculum topic on a range from 0 to 3, where 0 is for no capability, 1 for some capability, 2 for adequate capability and 3 for excellent capability. This will provide information on the university's strengths and weaknesses and help to determine which curriculum topics need to be developed to enable the university to design a NTM master's level programme that will be INMA compliant.

4.2.1.2. INMA preliminary programme description form

The purpose of the INMA preliminary programme description form is to provide information on the universities, faculties and departments involved in the development of the NTM programme and to describe the proposed NTM programme with respect to the following aspects:

(a) Policy, strategy and vision: It is important that the university and the department and/or faculty responsible for the delivery of the NTM programme has clearly stated policies and strategies that define the approach to delivering such a programme. These should be consistent with the national requirements for NTM education.

(b) Programme entry requirements: Nuclear educational programmes can be sustained through attracting the best students by successful marketing the programme. Scholarships can provide incentives for students to apply, but it is also important to establish which subjects and level of qualification are required to enter the programme.

(c) Programme delivery: Universities require excellent staff and facilities to deliver INMA-NTM programmes. These can be augmented by contributions from industry lectures and external facilities that may be available at the university. Experimental facilities can greatly enhance the teaching of the technology curriculum topics.

(d) Curricula: NTM curricula can vary due to local requirements, the design of the programme is based on mandatory and elective components and may also include soft skills such as communication. The programmes can also be delivered as full time or part-time, classroom based or distance learning, thus providing the flexibility required by students that are already employed.

(e) Quality control of the INMA-NTM programme: Quality control mechanisms are vital to ensure the continued success of any educational programme and to ensure that it remains relevant to the requirements of the stakeholders.

(f) National and international dimensions: The establishment of national and international collaboration and networks is used as a tool to exchange experts and information as well as to facilitate the full utilization of facilities. Networking of educational institutions has been widely recognized as a key strategy for capacity building and better use of available educational resources.

(g) Collaboration with industry on nuclear technology programmes: Close cooperation between industry and universities has been recognized by most countries as a vital factor in improving nuclear education and training,

The programme description form contains questions to guide the university to provide the relevant information. The preliminary self-assessment tool and programme description form are required to be submitted by the university to the IAEA prior to any mission that is conducted to assist a university with development of its NTM programme.

4.2.2. The INMA assist mission report

A report on the findings of the mission will be prepared by the mission team, which will be sent to the university and will typically consist of the following:

(1) Administrative information;
(2) Terms of reference;
(3) Details of the proposed university NTM programme, including meetings with university representatives and stakeholders;
(4) An inventory of existing university courses and their key elements that could be used for the NTM programme;
(5) An inventory of available equipment, facilities and human resources;
(6) Recommendations to the university on how to develop its INMA-NTM programme including addressing gaps not covered by existing courses;
(7) Appendices containing supporting information.

4.3. INMA ASSESSMENT MISSIONS

Universities and other educational organizations may request an INMA assessment mission to support the establishment of an INMA-NTM programme or, ideally, after a programme has been established. The mission determines whether the implementation of an NTM programme is INMA compliant and demonstrates the highest standards of professional conduct as befitting designation as an INMA member. The assessment, which is presented in the form of an assessment mission report prepared by the mission team, provides observations and recommendations for improvements as determined by the collective experience and interpretation of the team members.

The assessment process will normally include the following:

— An inventory of the NTM programme of the university, assessment of its gaps against the INMA curriculum topics, and discussion on proposed resolutions;
— An assessment and sharing of best practices relating to the implementation of INMA-NTM programmes, such as ensuring a high level of stakeholder involvement or effective utilization of IT tools to provide more accessible courses;
— An evaluation of the university's overall programme implementation and approach for its master's degree in NTM, and feedback on how to further improve or strengthen the programme, as well as INMA;
— Assistance, if appropriate, to the university to formulate the final detailed requirements and plans for implementation of their NTM programme.

4.3.1. Assessment information package

The assessment information package consists of the INMA-NTM programme self-assessment tool, programme description form, course description form and the courses delivery and student assessment description form, all shown in Appendix III.

4.3.1.1. INMA-NTM programme self-assessment tool

The INMA-NTM programme self-assessment tool is designed for universities to map their NTM programme against the INMA curriculum topics. Universities are requested to enter the number of hours delivered for each curriculum topic on each of the courses that comprise their NTM programme. Information is also required for each course on the breakdown of learning hours according to whether they are direct teaching, self-study or practical exercises, whether the direct teaching is in person, on-line or both; whether the teaching approach is theoretical, experiential or both; and how many hours of lectures are delivered by industry experts and leaders. This provides the mission team with comprehensive details on all the courses of the programme being assessed.

4.3.1.2. INMA-NTM programme description form

The purpose of the INMA-NTM programme description form is to provide an update on the INMA preliminary programme description form and to provide additional information on the outcomes of the university's NTM programme. Programme outcomes may be best defined as the quality and quantity of graduates, together with the roles and influence they fulfil in their careers and for their employers. An effective NTM programme should engage with organizations in the nuclear industry that employ their graduates to determine the employability of the students for a technology management career in the nuclear industry. Close collaboration with employers provides valuable feedback that leads to continuous improvement. Compared to the preliminary programme description form, extra information on the following topics should be included in the information package and discussed during the mission:

(a)   Industry demand for graduates with NTM education;
(b)   Graduation rate for the course or programme if data are available.

4.3.1.3. INMA course description form

The INMA course description form provides information on the aims of each course within the university's INMA-NTM programme and a description of each course with course subjects. Learning outcomes for each course are also provided and are described in the following under three dimensions: knowledge, demonstration and implementation:

• Knowledge of a subject requires remembering previously learned material, understanding the concepts and meaning of the material.

- Demonstration of the application of knowledge requires using learning in new and defined situations, understanding both the content and structure of the material. Demonstration is more reactive and is in response to training.
- Implementation of the knowledge requires formulating new structures from existing knowledge and skills, judging the value of material for a given purpose (i.e. knowing how and when to implement). Implementation is more proactive and implies the independent application of the ideas, concepts and methods taught.

4.3.1.4. INMA courses delivery and assessment description form

The courses delivery and assessment description form provides detailed information on the teaching methodology and assessment methods for each of the courses of the university's NTM programme.

4.3.2. The INMA assessment mission report

At the end of an INMA assessment mission, a summary list of main findings and recommendations will be offered to the university, which will then have the opportunity to review and provide comments on their content. The university should endeavour to return comments to the IAEA within an agreed time frame. The comments can then be incorporated into the assessment mission report to form the basis of any action plan, if required.

The assessment mission report will be prepared by the mission team. The team leader will coordinate the authoring of the report with the Scientific Secretary from the IAEA Nuclear Knowledge Management Section before submitting it to the university. The report should clearly address all the objectives stated in the terms of reference of the mission and document the team's findings and recommendations, including a summary of follow-up actions if required. The report should clearly present any gaps between the university's NTM programme and INMA-NTM compliance. The report will include contributions from each team member and summarize the team's main findings and conclusions, including a recommendation on whether or not the assessed programme has reached the standard required of an INMA-recognized NTM programme.

The structure of the assessment mission report is outlined below, together with details of the suggested content:

(1) Administrative information that includes project information relating to the INMA assessment mission and comprises:
    — Project number, if applicable;
    — Project title, if applicable;
    — Task title;
    — List of the mission team members;
    — Dates of the assignment;
    — Counterpart information (i.e. names and location);
    — Duty station location.
(2) Terms of reference describing the objectives of the INMA assessment mission, its scope and duties. Four separate subsections apply:
    — Objectives of the mission describing the objectives as agreed with the university;
    — Mission duties describing the form of the mission (i.e. how it was conducted);
    — Deliverables, including the assessment mission report produced as a result of this mission.
(3) A background providing information to the mission on the context of the status of NTM education and the issues facing the global nuclear industry and specific nuclear industry and NTM issues within the host country or university.
(4) A work programme outlining the programme of work undertaken during the mission with details of dates, times, locations and responsibilities. It can consist of the agreed agenda as prepared prior to the mission with any modifications, as appropriate.
(5) Assessment of the NTM programme and how the curriculum topics are taught and their key elements, gathered from presentations on the programme courses and discussions on the NTM programme self-assessment and information package.

(6) Recommendations to the IAEA for INMA from the university, the mission team or other parties involved with the mission. The recommendations to the IAEA may consist of the following:
  — Strategic IAEA initiatives that should be undertaken to support NTM educational programmes;
  — Recommendations on applicability of the good practices identified in NTM programmes for further enhancement within the INMA-NTM programme;
  — Other recommendations that are relevant to NTM education that can be implemented directly or be facilitated by the IAEA.
(7) Conclusions and recommendations to the university. These should aim to identify whether the university's existing NTM programme is INMA compliant, including good practices as well as areas that need to be developed. Wherever possible, the mission team will endeavour to provide pragmatic advice that can be translated into an action plan by the university at the end of the mission.

Typical conclusions and recommendations will cover one or more of the following issues:

  — Analysis as to whether the NTM programme includes the required 80% of the INMA curriculum topics;
  — Observations of good practice;
  — Strategic recommendations that may involve key stakeholders or supporting organizations;
  — General recommendations applicable to the university that relate to NTM education improvement;
  — Specific recommendations that could be applied to the university that relate to NTM education improvement and specifically to participation of the university in the INMA-NTM programme.
(8) Any references used in the report.
(9) List of appendices. The NTM programme self-assessment tools and assessment information package will be included as appendices together with other appendices as required that contain information, such as presentation summaries, lists of participants, self-assessment output, contact details and other information that is requested or provides value to the university.

The IAEA will send the assessment mission report with a cover letter to the university. The IAEA will restrict initial distribution to the university, the contributors to the report and relevant IAEA staff. Any further distribution will be at the discretion of the university or by agreement between the IAEA and the university.

The university will respond to the report and agree on the validity of the recommendations contained within the report. Once agreement has been reached, the university submits its response to the IAEA, including its strategy to address any issues identified in the NTM programme. It is expected that the university also commits to an action plan to address any gaps identified in the NTM programme.

4.4. UNIVERSITY NTM PROGRAMME ENDORSEMENT BY THE IAEA

If the Scientific Secretary of the IAEA and the mission team's recommendation is for the university to become a new INMA member, this decision will be conveyed to the Director of the Division of Planning, Information and Knowledge Management and the Nuclear Knowledge Management Section Head. If they concur that the programme is INMA compliant based on the conclusions of the assessment report and any required action plan, the IAEA can accept the recommendation for INMA membership. The university's programme may then be endorsed as an INMA-NTM programme and the IAEA shall confirm this through a formal letter sent to the university with an INMA-NTM programme certificate. Until such letter is received, the university should not assume that it is an INMA member.

For an INMA member to maintain its status and endorsement of its programme, a complete information package must be updated and submitted to the IAEA for review and evaluation every four years. Every eight years, the re-endorsement evaluation must include an assessment mission. All re-endorsements will be signified by a formal letter from the IAEA.

If it is concluded that an INMA member's NTM programme is no longer INMA compliant, the university is not meeting the highest standards of professional conduct as befitting an INMA member, that any agreed action plan has not been implemented satisfactorily, or for any other reason the IAEA deems sufficient, the INMA membership and programme endorsement may be revoked.

5. INMA WEB SITE

The INMA web site[2] offers the latest information about INMA, including information relating to the curriculum topics to all interested parties and organizations regardless of their membership. The information contained on the web site includes the following:

— A list of the universities and other educational institutions that are either INMA members or are interested in offering INMA-NTM programmes;
— The self-assessment tools, information packages, forms and report templates;
— Notice of upcoming INMA annual meetings;

INMA members are required to develop and maintain their own complete public INMA web site at their own institute. The INMA web site hosted by the IAEA will provide a link to these individual INMA member web sites. To provide a summary on the INMA web site of all the INMA-NTM programmes, INMA members are required to provide the IAEA with basic information updates on their current INMA-NTM course offerings, their official logo and a link to their INMA-NTM programme web sites.

6. INMA ANNUAL MEETINGS

The INMA annual meeting is hosted by either the IAEA or one of the INMA members. Attendees of the meeting may include INMA members, interested universities and other stakeholders. INMA members are expected to submit an annual status report and share their experience on implementing and delivering INMA-NTM programmes with other participants of the meeting. This provides valuable information on the status and development of INMA-NTM programmes including the endorsement process. The IAEA records the progress made by the universities in the implementation of their INMA-NTM programmes as well as any feedback and suggestions for further development based on discussions at the annual meeting.

7. CONCLUSION

NTM university educational programmes have been identified by both the nuclear industry and universities providing nuclear education as a mechanism to improve the safety, performance and efficiency of nuclear and radiological facilities and projects. The IAEA has established the International Nuclear Management Academy to support universities in the establishment of such programmes and to ensure that the nuclear industry is a valued and important stakeholder as the programmes are established and delivered. The breadth of the INMA curriculum topics ensures that the nuclear technology managers graduating from these programmes will have the appropriate skills, at the appropriate levels, to make significant contributions to enhancing safety, performance and efficiency at the facility and project level, as well as in the wider global nuclear community. This publication has been produced to help universities to establish INMA-NTM programmes that will be sustainable and to ensure continued positive development of the worldwide nuclear and radiological workforce.

---

[2] www.iaea.org/INMA

Appendix I

SUGGESTED COURSE SUBJECTS FOR EACH CURRICULUM TOPIC

To provide consistency in the curricula of INMA-NTM master's programmes suggested course subjects for each curriculum topic are presented in this appendix as a reference for universities and their INMA-NTM programmes.

1. CATEGORY 1: EXTERNAL ENVIRONMENT

The external environment category includes eleven curriculum topics and is presented below together with suggested course subjects. External in this context means external to the organizations that managers are working for. Some teaching elements for this category were derived from Ref. [4].

1.1. Energy production, distribution and markets — The global energy and nuclear energy environment in which nuclear organizations must remain competitive and efficient. The suggested teaching topics for this curriculum topic include the following:

— Global and local energy options and markets and the main issues, trends, and connection to international supply and demand for nuclear power;
— Energy distribution systems and their related geopolitical challenges, national choices for the operation of energy generation and distribution systems (peak/baseload);
— Electricity producing technology and characteristics in relation to the grid, especially with respect to reliability, cost and need for baseload following and backup generation;
— Infrastructure needed for electricity distribution grids, technical challenges to reliable grid design, lessons learned from grid failure events;
— Global economic drivers for the price of nuclear energy and nuclear fuel (general balance of costs, economics of the fuel cycles — mining, conversion, cost of electricity, cost of capital during construction, costs of outage, maintenance) as well as main controversies and point of focus (e.g. factoring or not greenhouse gas emissions market);
— Life cycle of nuclear energy installations and main nuclear energy projects and trends;
— Changing public perceptions of sustainable energy issues and options and the overall acceptance of nuclear power in this context.

1.2. International nuclear and radiological organizations — The purpose, roles, mission, goals, scope of operations, impacts, achievements, cooperation and interrelationships among key nuclear and/or radiological sector organizations, associations and stakeholder groups. This curriculum topic should provide a broad overview of the relevant nuclear and radiological sector organizations internationally, and it is suggested to include most of the following topics:

— International nuclear suppliers and design organizations (e.g. major reactor vendors);
— Non-governmental nuclear organizations;
— Important international nuclear associations and committees (e.g. United Nations Scientific Committee on the Effects of Atomic Radiation, Cooperation in Reactor Design Evaluation and Licensing, Multinational Design Evaluation Programme, Western European Nuclear Regulators Association, WANO);
— Major international nuclear and radiological organizations (IAEA, OECD/NEA, World Nuclear Association, International Commission on Radiological Protection, International Radiation Protection Association, United Nations Scientific Committee on the Effects of Atomic Radiation);
— Reactor owner's groups (e.g. CANDU Owners Group);
— Technical support organizations and networks (e.g. European Technical Safety Organisations Network);

— Nuclear networks, associations and societies (e.g. European Atomic Forum, European Nuclear Safety Regulators Group, Arab Network of Nuclear Regulators, Asian Nuclear Safety Network, Forum of Nuclear Regulatory Bodies in Africa, Ibero-American Forum of Radiological and Nuclear Regulatory Agencies, Heads of the European Radiological Protection Competent Authorities, European Nuclear Engineering Network, International Decommissioning Network);
— International nuclear research and development organizations and laboratories (e.g. Electric Power Research Institute, Halden Reactor Project);
— International non-nuclear organizations that interact with or have an impact on international or national nuclear organizations (e.g. United Nations, OECD/NEA, World Bank, International Monetary Fund, Greenpeace).
— Steps and drivers for establishing a national energy policy;

1.3. National nuclear technology policy, planning and politics — The policy aspects that are required for the successful and ongoing operation of nuclear and/or radiological facilities. Due to the long lifetimes of nuclear facilities, any possible changes in energy policy and the political environment over time should be recognized, and how these factors may affect the viable operation of nuclear facilities should be understood. The suggested teaching topics for this curriculum topic include the following:

— Governmental decisions and policies to establish nuclear as a generation source;
— Commitments to a nuclear programme (see Ref. [8]);
— Costs, timelines and challenges involved in such large scale commitments;
— Stakeholders and steps involved in the establishment of a national nuclear energy plan;
— Infrastructure needs (e.g. industrial, research and development) and existing experiences of countries developing or having developed a nuclear energy programme.

1.4. Nuclear standards — The existing international and domestic standards pertaining to the nuclear or radiological sector, including IAEA safety standards, other international standards and their influence on national regulatory requirements. The suggested teaching topics for this curriculum topic include the following:

— IAEA safety standards, other international standards (e.g. International Electrotechnical Commission, Institute of Electrical and Electronics Engineers, American National Standards Institute, American Society of Mechanical Engineers, Nuclear Information and Records Management Association, International Organization for Standardization) and the scope and guidance they provide;
— Existing role, nature, challenges and benefits of nuclear standards applicable across the nuclear sector;
— Influence of international standards on national regulatory requirements, national standards and the design, licensing, and operation of nuclear facilities in general;
— Consideration of nuclear standards in the national context (e.g. proper regulatory and legal interpretation of the hierarchy of national standards, regulatory requirements and international standards within national context).

1.5. Nuclear and radiological law — The framework of national legal obligations, international conventions, treaties, agreements and United Nations Security Council resolutions that establish the environment in which nuclear organizations and their managers must function. The suggested teaching topics for this curriculum topic include the following:

— Fundamentals of law (e.g. basic principles, origins, main approaches to, types of law);
— Government laws, decrees, treaties and regulations;
— Differences between different kinds and hierarchy of law (e.g. private versus company law);
— Major international treaties, international instruments and main references relating to nuclear (e.g. Convention of Nuclear Safety, Joint Convention on the Safety of Spent Fuel Management and on the Safety of Radioactive Waste Management, IAEA safety standards, Convention on Early Notification of a Nuclear Accident);
— Importance of a proactive attitude towards legal aspects and features (e.g. knowing when to seek legal advice on matters of nuclear or radiological law);
— Laws directly and indirectly linked to the nuclear and radiological sectors (objectives and principles of legislative processes associated with nuclear law);

— IAEA publications on legal framework for nuclear safety, security and safeguards [9–11];
— Legal aspects involved in nuclear managers' decisions and concrete case examples (e.g. regulatory authority and legal obligations of nuclear facility licence holder, rights of individual, radiation and nuclear safety, nuclear security, environmental protection, public health and safety, emergency response, labour health and safety, criminal law, rights of individuals, transport law);
— Typical government laws and decrees relating to nuclear facilities and nuclear activities in general;
— Relevant legal terminology, laws and decrees relating to nuclear liability.

1.6. Business law and contract management — The legal context, including the liability and contractual issues and their management, taking into consideration lessons learned from previous legal interpretations and the enforceability of contracts. Resolution management is required for the successful implementation of some contracts. The suggested teaching topics for this curriculum topic include the following:

— Basic principles of business law and contract law;
— Terminology and expressions used in business and contract law;
— Different types of contracts for different nuclear projects (new build, refurbishment, power uprate, decommissioning), typical requirements and rationale (needs, purposes);
— Lessons from major project contracts in the past for the nuclear sector;
— Stakeholders who may be directly (e.g. vendor or constructor) or indirectly (e.g. international organizations, regulators, sub-contractors or banks) impacted by contracts;
— Various contract roles and financial or liability positions and the evaluation of risk of different stakeholders and different types of risk in regulatory, finance, construction;
— Different steps of legal agreements, informal agreements and links with procurement or competitive tendering processes;
— Payment, risk and scheduling considerations for contracts and contract negotiations, including incorporation of penalties and potential impacts on project scheduling and quality;
— Special terms and conditions to contracts (e.g. legal basis, insurance, intellectual property related aspects, unexpected force majeure events);
— Conflict resolution and management and claim negotiations.

1.7. Intellectual property management — Intellectual property rights and their management, as well as associated issues, for nuclear facilities and projects across the nuclear and radiological sectors. The suggested teaching topics for this curriculum topic include the following:

— Intellectual property rights (e.g. patents, copyright, design rights trademarks);
— Organizational approaches and perspectives for managing intellectual property rights;
— Intellectual property strategies (including when to register an intellectual property provision in contracts and other types of agreements);
— Intellectual property related considerations in decisions (e.g. legal considerations, knowing when and where to look for intellectual property issues, risks);
— Different kinds of intellectual property rights (registered and unregistered, licensed intellectual property, ownership versus right to use) and related issues;
— International and local approaches to intellectual property protection;
— Intellectual property transfer issues and risks in new build projects (e.g. restricted access to vendor's protected design information, intellectual property transfer, scope of design basis documentation handover);
— Role of intellectual property in relation to safety (e.g. counterfeit parts);

1.8. Nuclear and radiological licensing, licensing basis and regulatory processes — The regulatory licensing processes for nuclear and/or radiological facilities and related processes and practices. The suggested teaching topics for this curriculum topic include the following:

— Fundamental principles for licensing (e.g. definition of licensing for nuclear installations, approaches for licensing in various countries, role of regulator);

— Licensing procedures of nuclear or nuclear related facilities (oversight inspections);
— Stages for the licensing process and international approaches (e.g. IAEA milestones [8]);
— Objectives, scope and content of major licence submittals (e.g. preliminary safety review, final safety review, periodic safety review assessments);
— Regulatory body's processes and practices (e.g. assessments, authorizations, inspections, enforcements, development of regulations and guides, public awareness);
— Risks associated with licensing procedures (such as risks associated with finance, scheduling, cost, credibility, lack of common international approach and technical basis);
— Periodic relicensing versus fixed term;
— Licensing for life extension of a nuclear power plant.

1.9. Nuclear security — The national and international issues, frameworks, norms, obligations and approaches relating to physical nuclear security (the prevention and detection of, and response to, theft, sabotage, unauthorized access, illegal transfer or other malicious acts involving nuclear material, other radioactive substances as well as their practical implementation and impact on licensed nuclear or associated facilities. The suggested teaching topics for this curriculum topic include the following:

— Issues, environment and trends governing nuclear security (e.g. implications of global threats, increased relationship with safety).
— Existing global legal framework for security (Convention on Physical Protection of Nuclear Material [12]).
— International and national sources (e.g. Ref. [13]), instruments and plans for nuclear security (e.g. national security regime, plan and systems; cooperation, relevant organizations).
— Relevant national institutions and basic security principles, concepts and terminology involved in physical protection programmes.
— Risks of unauthorized removal with the intent to construct a nuclear explosive device, risk of unauthorized removal, which could lead to subsequent dispersal, and risk of sabotage should be taken into consideration for the protection of nuclear material and nuclear facilities (e.g. nuclear power plants, fuel fabrication facilities, waste facilities, research reactors); insider threats should also be understood.
— Organizational approaches (e.g. best practices, main differences, lessons learned and information exchange collaborations) and responsibilities for physical protection of facilities (both civil and other, notably within the security plan).
— Organizational approaches and responsibilities for physical protection during transport of nuclear material (both civil and other) [14].
— Approaches for devising security strategies at various levels (e.g. threat assessments, design basis threats, defence in depth, scenario analysis, internal analysis, graded approaches) [15].
— Approaches to establishing emergency (contingency) measures, various instances of implementation and key elements including the combination of threats (e.g. safety threats, followed by security related threats and vice versa).
— Need for a security culture and its interplay with and consequences on safety and safety culture and safeguards (sometimes complementary, sometimes contradictory, such as delaying intruders versus minimum delay time for emergency vehicles to access the buildings) [16].
— Duty and liabilities for managers in various situations.
— On-site measures implemented and both internal and external response forces, and typical interfaces to prevent, manage and investigate security related incidents (nuclear security forensics).
— Actual instances and existing experience, learning from case studies of past events.
— Necessity of comprehensive risk management and quality assurance processes.
— Control of access to sensitive security information and related threats including cybersecurity threats and measures [17].

1.10. Nuclear safeguards — The national and international issues, frameworks, norms, obligations and approaches relating to nuclear safeguards as well as their practical implementation and impact on licensed nuclear facilities. The suggested teaching topics for this curriculum topic include the following:

— Past, present and projected international treaties, agreements and protocols in the field of nuclear safeguards (e.g. Treaty on the Non-Proliferation of Nuclear Weapons) origins and achievements;
— State system of accounting for, and control of, nuclear materials;
— Complementary measures for non-proliferation including the bilateral and multilateral approach, their goals and achievements but also their impact on industry (inter alia increased credibility for proactive states);
— Concepts, origins and issues relating to safeguards, instruments and export control;
— Past and present issues and lessons relating to proliferation;
— Principles and practices involved in safeguards, including inspections;
— Importance of safeguards and non-proliferation measures and instruments and their provisions to the nuclear industry in general, and more specifically to nuclear facilities;
— Control of access to sensitive safeguards information and related threats;
— Duty and liabilities for managers in various situations;
— Actual instances and existing experience;
— Assessing safeguards with respect to a specific nuclear technology;
— The inclusion of safeguards in nuclear reactor and nuclear facility design and construction [18, 19];
— The IAEA's role and reference to available training and guidance materials.

1.11. Transport of nuclear goods and materials — The international and domestic frameworks for the safe transport of nuclear goods and materials, and the safety and security risk associated with their transport. The suggested teaching topics for this curriculum topic include the following:

— International instruments governing the transport of nuclear material, such as the Convention on Physical Protection of Nuclear Material [12];
— Organizations involved and related safety and risk processes [20];
— Features relating to secure transportation of nuclear material [13, 14];
— Categorization of radioactive waste and materials and respective requirements (e.g. packaging types);
— Legal provisions and actors governing international, interregional and intermodal transportation (e.g. International Maritime Organization, International Civil Aviation Organization, Intergovernmental Organisation for International Carriage by Rail, United Nation Economic Commission for Europe);
— Roles, perspectives and actions of the various stakeholders (e.g. states, operators, civil society, activists) in normal and abnormal situations (e.g. emergencies during transport);
— Early notification of incident schemes;
— Key technologies that support safety and security in the transport of nuclear goods and materials.

2.   CATEGORY 2: TECHNOLOGY

The technology category includes fifteen curriculum topics and is presented below together with suggested course subjects.

2.1. Nuclear or radiological facility[3] design principles — The technology, management, safety, security and non-proliferation influences on the principles of designing a nuclear power plant or other major nuclear or radiological facility. The suggested teaching topics for this curriculum topic include the following:

— Approaches and rules for safe design (e.g. defence in depth and diversity), systemic safety approaches;
— Concepts of design and operation of nuclear reactors;

---

[3] Different university INMA-NTM programmes may choose whether to focus on nuclear power plants or other licensed facilities and select their teaching elements accordingly.

— Rules and technical criteria and codes to be followed during design (e.g. emergency core cooling system standards, American Society of Mechanical Engineers standards, the French code RCC-M (Règles de conception et de construction des matériels mécaniques des îlots nucléaires réacteurs à eau sous pression), Design and construction rules for mechanical equipment in PWR nuclear islands, Japan Electric Association guidelines and codes, IAEA standards and guides [15, 18, 19, 21], national standards);
— Using different types of simulator (e.g. full-scope basis principles reactivity dynamics, plant modes, reactor trip);
— Ergonomics applied to the design phase and human factor specific considerations;
— Principles and methods of radiation protection used in design;
— Scope, content and depth of design at various project phases (e.g. conceptual versus detailed);
— Quality management of design (e.g. design verification, control of interfaces, design review);
— Relevant design methodologies and importance of safety analysis and validation;
— Specifics in the principles of designing a nuclear or radiological facility, for example, special considerations for nuclear safety, radiation safety, non-proliferation, nuclear security, criticality management (suggest addressing the following: seismic criteria, safety systems, external hazards, design basis accidents, severe accident conditions, post-accident monitoring, emergency preparedness);
— Design features of a nuclear power plant or other major nuclear or radiological facility technology (suggest highlighting the safety feature differences between reactor generations).

2.2. Nuclear or radiological facility[3] operational systems — All major plant systems and operating modes of a typical nuclear or radiological facility as well as the main components of each facility or system, and their functions. The suggested teaching topics for this curriculum topic include the following:

— Nuclear reactor system functions and power plant technologies;
— Basics of plant layout, plant operating states, plant technical systems, equipment functioning and main operating processes;
— Primary systems (e.g. description, component functioning, operating modes in normal and abnormal situations);
— Appropriate plant operating modes and corresponding systems interrelations;
— Overall comprehension of instrumentation and control systems to ensure safe operation;
— Requirements and safety importance of routine status checks of equipment and systems, effective monitoring, attention to detail with mechanical and electrical devices important to safety;
— Operating team's overall responsibility for the safety of all workers on-site at all times (e.g. nuclear safety, plant security, firefighting);
— Operating states, modes, technical systems, equipment, team's responsibility, and key parameters (e.g. start-up, shutdown, power changes, load following);
— Necessity to maintain appropriate attitudes making useful, comprehensible notes of malfunctions and leaks, pre-diagnosing their causes and contributing to fixing them;
— Highlighting the key differences in reactor technologies (e.g. boiling water reactor, pressurized water reactor, pressurized heavy water reactor, and fast reactors).

2.3. Nuclear or radiological facility[3] life management — The principles of long term operation and management of plant life. Nuclear facilities have special long term operational needs, given the level of their safety, reliability, economics and regulatory licence requirements. The suggested teaching topics for this curriculum topic include the following:

— Assessment of design margins, taking into account all known ageing and wear mechanisms and potential degradation in normal operation, including the effects of testing and maintenance processes;
— Provisions for monitoring, testing, sampling and inspecting safety related equipment to assess ageing mechanisms, verify predictions, and identify unanticipated behaviours or degradation that may occur during operation as a result of ageing and wear;
— Predictive and failure analysis for equipment maintenance;
— Overall functioning of on-line sensors and monitoring devices;
— How to use analysis of real time asset performance data for decision making;

— Understanding that the foundation of the asset management policy is that nuclear safety is the overriding priority, although plant performance and economics is a key factor in achieving this;
— Strategies for the timing of major equipment replacement (and related trade-offs), proactive investment in maintenance, major equipment replacement decisions, life extension, and refurbishment;
— Equipment that is required for nuclear safety and understanding how its performance is monitored;
— Selecting the preventive maintenance programmes and the life cycle management strategies to be adopted to optimize reliability and availability; this is done at both system and component levels, and is used to anticipate and manage ageing and degradation effects that affect nuclear safety and reliability;
— Equipment performance data collection and its use in the process on improving equipment reliability and availability, resulting in increased nuclear safety margins;
— Interactions between a corrective action programme and maintenance programme;
— Configuration management, defined as the process of identifying and documenting the current state and characteristics of a facility's structures, systems and components (including computer systems and software), and of ensuring that changes to these states and characteristics are properly developed, assessed, approved, issued, implemented, verified, recorded and incorporated into the facility documentation;
— IT's role in assisting the management of power companies as well as increasing the efficiencies of plant design, operation and maintenance work more than ever (e.g. suggest providing examples such as enterprise application software, enterprise resource planning systems, parts and supplier management, 3-D computer aided design models, outage planning and work management systems);
— IT's role as an enhanced performance tool across the board from design to construction planning, maintenance or training scheduling activities; integrated management systems can only exist today through IT.

2.4. Nuclear or radiological facility[3] maintenance processes and programmes — The strategies and methodologies that may be used to establish maintenance programmes at nuclear or radiological facilities, with various levels of complexity to meet licence conditions and design specifications. Planning and activity scheduling as a management tool is also required for an effective quality assurance programme. The suggested teaching topics for this curriculum topic include the following:

— Regulatory elements and norms relating to nuclear maintenance activities [22];
— Maintenance work performed within a regulatory framework, which takes into account high quality requirements while ensuring safety of the installation and worker security by systematically applying as low as reasonably achievable (ALARA) principles and environmental protection;
— Different methodologies to establish maintenance programmes;
— Complexity of maintenance work in nuclear or radiological facilities that need to use planning and activity scheduling as a management tool, and an effective quality surveillance programme;
— Appropriate level of detailed maintenance work instruction so that workers, schedulers, and other affected organizations can carry out the activities in a planned and controlled manner (e.g. work package approvals, equipment status monitoring, foreign material exclusion programmes, pre/post job briefings, equipment tag-outs, control room communications and coordination of field work);
— Process for coordination of integrated discipline of maintenance work packages to ensure involvement of the appropriate persons and the proper sequence of carrying out the work;
— How feedback and history from previous maintenance works are recorded and used in the planning process (e.g. deficiency reports);
— How to describe types of maintenance (e.g. corrective, predictive, preventive, reliability centred) that can be used by nuclear facilities, their definitions, and their applicability;
— Process to identify, order, receive, store, and install proper parts and materials for work activities while meeting all quality requirements, and how safety related parts and components are properly controlled, segregated, identified, and issued in all material storage areas; appropriate unused parts and materials promptly returned to inventory.

2.5. Systems engineering for nuclear or radiological facilities — The interdisciplinary approach enabling the successful design and implementation of complex systems. This includes defining the required functionality early in the development life cycle, documenting requirements and then proceeding with design synthesis and system validation, which delivers successful implementation (i.e. to achieve equipment system reliability, safety, performance, economics or other goals). An understanding is required of how system engineering plays an important role throughout the life cycle of the nuclear or radiological facility. The suggested teaching topics for this curriculum topic include the following:

— Contributing to creating consensus regarding the problem definition and the corresponding technical requirements;
— Describing, identifying or defining desired outcomes and success criteria;
— Identifying constraints and the expected system environment;
— The purpose of having a safety margin philosophy to address the expected operating conditions and environment and how that leads to design guidance (concepts of limits and margins, such as operating margin, design margin, analytical margin);
— Appropriate regulatory requirements and international standards issued by the IAEA and other international organizations;
— Selection of a technical solution to balance technical and non-technical features of the system;
— Documentation of design requirements, design rationale and basis, conceptual, general and detail design (i.e. types and levels of design information and its representation, such as the system architecture, subsystem design, component level design and interfaces, other integrated design information and documentation needed across systems and subsystems);
— The important role of systems engineering during the operational phase of nuclear and radiological facilities;
— Management of stakeholder expectations throughout the project life cycle;
— Conversion of functional and behavioural expectations into technical terms with performance requirements and corresponding tracking indicators;
— Planning and prioritizing activities for technical teams; need for cross functional coordination;
— Implementing proper knowledge and design quality management strategies that provide integration of technical knowledge and information from reports, trend analyses and lessons into a knowledge management system that will enable proactive information use, assist in problem solving and improve decision making.

2.6. Nuclear safety principles and analysis — The nuclear safety fundamentals, their principles, analysis methods and how the industry is responding to the safety requirements. The suggested teaching topics for this curriculum topic include the following:

— Safety fundamentals from conception and design to operations and dismantling (e.g. design of nuclear power plants for safety concerns, containment barriers, defence in depth, safety functions);
— Regulatory requirements within a historical perspective and current trends to link safety related methods with major accidents, risk control innovation and security issues;
— Reports addressing the fulfilment of safety requirements during a nuclear power plant life cycle (e.g. safety report, risk control analysis);
— The IAEA system of international safety standards (i.e. IAEA Safety Standards Series No. GSR Part 4 (Rev. 1), Safety Assessment for Facilities and Activities [23]) and its application in different countries;
— Safety elements needed for reactor criticality control, radiation exposure reduction, effective containment and emergency evacuation;
— Safety elements in fuel front end factories, spent fuel handling and waste treatment facilities;
— Safety analysis methodology (probabilistic safety assessment, probabilistic risk assessment, probabilistic and deterministic studies), understanding human and organization factors impacting safety;
— Effect of safety requirements on nuclear power plant general operating rules, including the role of the safety engineer.

2.7. Radiological safety and protection — The justification and optimization of protection for planned, emergency and existing exposure situations. Safety and protection are ensured by the understanding of the radiological impact on different materials and on the human body. Substantial control on each nuclear site is maintained by adherence to the radiological regulations in each country. The suggested teaching topics for this curriculum topic include the following:

— Radiation protection fundamentals based on a global knowledge of the scientific fundamentals of radiation and the interaction of radiation with matter;
— Ionizing radiation effects and their risk to humans (non-stochastic (deterministic) or stochastic effects);
— Potential exposure pathways for discharges to atmospheric and aquatic environments, and soils or sediment;
— Appropriate regulatory requirements and international standards issued by the IAEA and other international organizations (especially including approaches to authorized and optimized discharge limits);
— Health and biological effects of radiation and know-how to apply reduction risk methods (ALARA principle) for radiation and contamination risks;
— Radiation sources and their isotopic inventory for effective protection;
— Protection methods from calculation to protection material selection;
— Basic knowledge of source utilization for radiation measures, and safety and health impact requirements.

2.8. Nuclear reactor physics and reactivity management — The general awareness of the physical principles needed for the design, construction and operation of nuclear reactors. This also includes all the factors that affect reactivity and criticality. The suggested teaching topics for this curriculum topic include the following:

— Reactor physics fundamentals, core characteristics over core life and how reactivity control systems operate for effective control of core reactivity during normal, abnormal and emergency operating conditions;
— Principles of design and construction of nuclear reactors, the main reactor types and their features;
— In-core poison behaviours with core age and the necessity to control their concentrations in any power transient;
— Understanding the effects of positive or negative coefficient of reactivity;
— Core reactivity coefficients variation with core life and the corresponding reactor control measures, including related databases and other existing tools;
— Essential reactor and plant indications to monitor core reactivity;
— Human performance error reduction tools for changes to core reactivity;
— On-load and off-load refuelling;
— Risk of critical excursion inside the different installations of the nuclear fuel cycle, the associated safety arrangements, the calculation methods and all major past reactivity accidents;
— Criticality safety of the transport of fissile material.

2.9. Nuclear fuel cycle technologies — The entire fuel cycle, from mining to final disposal. It includes all the technologies used to produce and manage nuclear materials as well as the security of the fuel supply, reduction in fuel cycle costs, management of the waste streams, and non-proliferation issues. The suggested teaching topics for this curriculum topic include the following:

— Different stages of the fuel cycle from uranium extraction to fabrication of fuel assembly, optimization of fuel burnup, reprocessing to utilize the spent fuel, reduction and treatment of the wastes arising;
— Materials used, their supply chain and strategic issues and role in defence in depth approaches;
— Overall in-core fuel management;
— Specifics of fuel management for a reactor fleet;
— Closed versus open fuel cycles and possible reuse of nuclear materials;
— Concepts of the different possible options (storage and recycling), their management, the industrial logic for optimization and economic aspects;
— Principles that guide the choice of the processes;
— Chemistry needed for the various stages of the cycle, related supply chain and the corresponding risk prevention requirements;
— Chemical mechanisms and geochemistry, which govern the evolution of storage;
— Different stakeholders at each stage of the cycle and the best method of interaction.

2.10. Radioactive waste management and disposal — The technologies associated with radioactive waste management including the stringent controls on radiological releases (solids, liquids, airborne materials, gases) to the environment. It ensures all measures are adhered to regarding the safety of people and the environment. The suggested teaching topics for this curriculum topic include the following:

— Radioactive waste management, with an emphasis on the stringent controls on plant releases (liquid, airborne materials, gases) to the environment;
— Different types of radioactive wastes, their disposal routes (and the economic aspects of these processes) including suitable means of controlling releases in a manner that conforms to the ALARA principle;
— How the plant ensures the necessary retention factors under the expected prevailing conditions of any releases through reliable and effective filtration systems;
— Rules and regulations associated with waste management and waste transport that foster implementation of corresponding radiation protection systems;
— Systems, specification for transportation material (casks), specific legislations and international agreements;
— Economic aspects of all these processes.

2.11. Nuclear or radiological facility decommissioning — The technological challenges and regulatory aspects associated with the end of the operating life of a nuclear or radiological facility. Consideration should be given to the various decommissioning strategies that have alternative financial and radiological implications. The suggested teaching topics for this curriculum topic include the following:

— Radioactive waste management with an emphasis on long term waste management routes whose availability is a condition for the success of decommissioning programmes;
— Different types of radioactive wastes, their disposal routes (and the relative economic aspects of these processes compared to other activities);
— Different stakeholders involved, perspectives (risk) and means to better communication;
— Analysis and resolution of problems associated with the decommissioning of nuclear facilities;
— Appropriate decommissioning specific rules and regulations;
— Radiation protection associated with waste management and decommissioning;
— Waste and effluent transport, as well as the application of specific related concepts such as the best available technique and best environmental practice.

2.12. Environmental protection, monitoring and remediation — The hydrological and ecological effect of nuclear and radiological facilities during normal operation through to decommissioning, and from the consequence of accidents on the local and wider environment. This curriculum topic also covers the effect of contamination of the local environment from radionuclides due to historical discharges, accidental releases, as well as releases from non-nuclear facilities such as mining, oil, gas and medical industries, and any subsequent environmental remediation. The suggested teaching topics for this curriculum topic include the following:

— Possible hydrological effect of nuclear power plants, including hydrometeorological impacts of cooling towers and effect of discharge in receiving systems;
— How to monitor hydrological and ecological effects, potential improving features, best practices;
— Possible consequences of nuclear accidents on ecosystems over different time periods and existing accumulated experience;
— Thermal loading and the related potential ecological effect;
— Radionuclide transfer in dissolved, particulate and sedimentary forms;
— Problems relating to successive power plants along the same river and various cooling sources;
— Possible ecological threats to nuclear power plant operations (e.g. comb jellies, algae, zebra mussels, calcium deposits);
— Computer modelling of near shore currents for nuclear power plant design, and of contamination dispersion forecasts (and their limits).

2.13. Nuclear research and development and innovation management — The technical innovation processes in nuclear and radiological sectors and industry research and development. This includes familiarization and knowledge of the role and scope of nuclear research and development organizations and their specific challenges. Knowledge of the current trends in next generation nuclear technologies provides an understanding of the current timeline for the deployment of various innovations. The suggested teaching topics for this curriculum topic include the following:

— Role of research in the nuclear and radiological sectors and its short and longer term effect (including spill over beyond the sectors). Importance of nuclear and radiological research facilities (e.g. research reactors), the various historical strategies and investments in nuclear science that led to the development of commercial nuclear power applications should be addressed.
— Management of research and development processes, innovation, and technology development.
— Emerging reactor technologies and major players involved (e.g. Generation IV technology and beyond).
— National, regional and international efforts in research and development (programmes, financing, cooperation, institutions such as the Joint Research Centre and the ITER International Fusion Energy Organization, fast reactor development).
— How the power output of current nuclear power plants has been increased through innovations in turbine design modifications, chemistry management and the addition of digital instrumentation and control systems.
— How, by using 3-D computer aided, design based plant information models and modular construction technology, the average project construction schedule duration has decreased significantly.
— Range and types of computer codes for design analysis and validation, the analysis of the performance of nuclear power plants (e.g. fuel behaviour codes, thermohydraulic codes).

2.14. Applications of nuclear science — The range of applications of nuclear science in the fields of research, medicine, industry, food and agriculture, the environment, security and space exploration. The suggested teaching topics for this curriculum topic include the following:

(a) Medical applications — Deals with the medical application of nuclear technologies such as radioisotopes and accelerators. The aim is to be aware of both medical risks and benefits of radiological exposure and to be aware of medical applications such as nuclear imaging for cancer diagnosis and nuclear medicine as a basis for cancer therapy and treatment, including the following:
   (i) Concept for the production (research reactors, accelerators) of radioisotopes and the status and distribution of these sources (market for radioisotopes and prices) as well as the potential effects of a reduction in the production of isotopes due to a temporary or permanent closure of a research reactor;
   (ii) Medical consequences of types of radiological exposure and the basic approach and principles behind treatment for each technology for diagnostic purposes;
   (iii) Concepts underlying the use of radioisotopes in nuclear radiotherapy;
   (iv) Concepts underlying the use of radioisotopes in nuclear oncology;
   (v) Existing research and development programmes in the field and associated main innovations.
(b) Nuclear food and agriculture applications — Deals with nuclear science applied to food and agriculture. The aim is to be able to play an ambassador role for the peaceful use of nuclear science in food and agriculture such as food irradiation, pest control, fertilizer and their safety issues, including the following:
   (i) Concepts underlying the use of radioisotopes in agriculture for pest control technologies (tsetse fly, fruit fly), mutation induced, drought-resistant food plants, and disease-resistant crops;
   (ii) How researchers use radioisotopes to learn when to apply fertilizer, and how much to use, preventing overuse and reducing a source of soil and water pollution;
   (iii) International collaboration (e.g. Food and Agricultural Organization of the United Nations) to face current challenges;
   (iv) Concepts and implementation underlying the use of radioisotopes in food irradiation to kill bacteria, insects and parasites that can cause food-borne diseases (salmonella, trichinosis) as well as to delay spoilage;
   (v) Scale of production and providers of isotopes used in food irradiation;
   (vi) Food irradiation in the light of proliferation, safety and security.

(c) Industrial and other applications of nuclear science — Deals with nuclear science applied to industrial processes such as non-destructive testing (for the quality of goods, and industrial and civil infrastructures), security, forensics and environmental clean-up and the use of nuclear isotopes and accelerators. The aim is to be aware of the peaceful use of nuclear science in fields outside power generation, including the following:

(i) Concepts underlying the use of radioisotopes in industry to improve the quality of goods; to develop highly sensitive gauges to measure the thickness and density of many materials; to use devices to inspect finished goods, metal parts and welds for weaknesses and flaws (non-destructive techniques); to track leakage from piping systems and monitor the rate of engine wear, equipment corrosion and filtration systems; to generate heat or power for remote weather stations or space satellites (i.e. a nuclear battery);

(ii) Concepts underlying the use of radioisotopes in many consumer products from smoke detectors to photocopiers, and from watches to cosmetics as well as for the determination of water and soil properties in hydrological investigations including snow gauging, soil moisture and density determinations, measurement of suspended sediment concentrations in natural streams, and nuclear well logging for groundwater exploitation;

(iii) Concepts underlying the use of radioisotopes in forensics and in other research fields needing material dating.

2.15. Thermohydraulics — The general knowledge of the physical principles needed to ensure adequate cooling of nuclear fuel during operation, shutdown, accident conditions and long term storage. This includes cooling of containment and consideration of ultimate heat sinks. The suggested teaching topics for this curriculum topic include the following:

— Principles of heat transfer including heat conduction, convection, and radiative heat transfer;
— Principles of natural convection, forced convection, laminar and turbulent flow, boiling and condensation/multiphase flow;
— Relationships for boiling crisis/critical heat flux, and the impact on nuclear safety;
— Basic principles for calculating/modelling thermohydraulic processes;
— Thermal physical properties of possible coolants;
— Basic design principle in the selection of the appropriate coolants for the reactor core, fuel storage applications, conventional building cooling, heating, ventilation and air conditioning, emergency cooling systems;
— Design of heat transport systems and establishing heat sink;
— Impact of thermohydraulics on reactor core physics;
— Monitoring of thermohydraulic parameters and safety margins;
— Basic principles of thermohydraulic behaviour during design basis accidents and severe accidents;
— Thermohydraulics of containment processes during accidents.

3. CATEGORY 3: MANAGEMENT

The management category includes eighteen curriculum topics and is presented below together with suggested teaching topics.

3.1. Nuclear engineering project management — The management of construction, refurbishment and decommissioning projects. This involves the various specifics of small, medium and large design and build projects and related procurement. Key aspects of any nuclear project include decision making processes, worker qualification, planning, project monitoring and control, documentation and communication management. Risk management, supplier quality control, licensing processes and major stakeholder roles and responsibilities should also be included. The suggested teaching topics for this curriculum topic include the following:

— Roles, functions, responsibilities and challenges of project managers;
— How to identify project non-statutory stakeholders;
— How to identify project statutory stakeholders (owners, partners, sub-contractors);

— Interfaces and related risks (notably communication) in various nuclear project configurations (e.g. various partners, cultures, legal environment, politics, local resources);
— Legal arrangements and influence on nuclear projects (contract type and strategy);
— Key drivers and phases for any nuclear projects (e.g. decision, training, pre-planning, planning, scheduling, documenting, organizing, monitoring, controlling), as well as the main differences between planning and implementing;
— Scheduling issues and major risks in nuclear projects;
— Issues including procurement, supplier, safety, risk and regulation (economic, safety) arising from delays in planning, cost of non-compliance with regulation;
— Good planning assumptions and controlling project execution sequence, dependencies;
— Good project management integration ensuring objective harmonization and cooperative business relations;
— Project planning tools and typical practice in nuclear projects (e.g. Gantt charts, project work breakdown structure, resource levelling, material management);
— Budgeting practices and issues;
— Human resource related requirements as main success factors (e.g. recruitment, worker qualifications, training, teamwork, delegation, staff motivation, negotiation and decision making);
— Key processes, such as safety related and quality control processes (e.g. supply chain management, configuration management, inspections, record keeping);
— How to monitor project performance and key drivers (e.g. performance metrics, relevant indicators, monitor project schedule and cost performance);
— Risk and approaches to risk management in nuclear new build projects;
— IT infrastructures needed and expected to enable effective management control of work processes, document control and the revisions process, control of communications, schedule management;
— In order to act as an intelligent customer, being aware of the competencies, best practices and tools to effectively manage performance and quality assessment of the vendor's work scope, including monitoring interfaces with subcontractors, regulatory processes;
— Main objectives and influencing factors to create a healthy organizational culture with a positive work environment and labour relations;
— Project management, providing leadership to control the activities of many technically based groups without having direct line authority over them.

3.2. Management systems in nuclear or radiological organizations — Various management systems in nuclear or radiological organizations include, for example, an operations management system, a training management system, a supplier management system, a quality management system, a work management system, an outage planning system and a licensing or regulatory compliance system. This curriculum topic emphasizes the importance of management systems to ensure that work processes are planned, monitored and controlled in a safe and systematic manner. It also acknowledges the importance of an integrated approach. The suggested teaching topics for this curriculum topic include the following:

— Management system coordinating all elements of an organization into a coherent framework: structure, resources (including knowledge), processes, personnel (including core competencies), equipment, organizational culture, documented policies and processes.
— Top level structure of management system including:
  • Vision, mission and goals of the organization;
  • Policy statements of the organization;
  • Organizational structure;
  • Levels of authority, responsibilities and accountabilities of senior management and organizational units.
— Structure of the management system documentation.
— Overview of the organization's processes.
— Responsibilities of the process owner.
— Arrangements for measuring and assessing the effectiveness of the management system.
— Integration of safety, health, environmental, security, quality, human and organizational factors, societal and economic elements and graded approach to the management system are encouraged.

3.3. Management of employee relations in nuclear or radiological organizations — The aspects to ensure a collaborative and cooperative relationship with and between employees, including contractual (i.e. collective labour agreement) considerations, emotional and trust issues, physical and practical aspects. It includes consideration of how employee relations may be influenced by local employment laws and cultural norms. A key focus is how to create an appropriate work environment that supports the safety and economic objectives of the organization. Factors of organizational culture, accountability and workforce performance should be addressed (e.g. establishing trust through open communications and a supportive culture of knowledge sharing and promoting positive gender relations). The suggested teaching topics for this curriculum topic include the following:

— The role and influence of employee relationships and the corresponding effect on organizational performance;
— Factors affecting employee engagement and job satisfaction and the benefits of a stable workforce;
— Factors that influence organizational culture and the work environment on labour relations;
— Collective bargaining processes, unionized labour and labour law and its effect on the approach to management of nuclear organizations;
— The manager's role in creating and maintaining a positive working environment, promotion of dialogue and open communication, and keeping employee engagement and motivation high in the organization.

3.4. Organizational human resource management and development — The aspects of the employment cycle. This curriculum topic also includes performance evaluation systems, workforce planning and adjustments, and ensuring the workforce can adequately meet its human resource responsibilities. Other factors to be addressed include succession planning, salary and benefits administration and organizational structure and performance. The suggested teaching topics for this curriculum topic include the following:

— Ensuring that individuals have the competencies needed to perform their assigned tasks, organizing work effectively, anticipating human resources needs, and monitoring and continually improving performance. Competency in this context is the ability to put the skills, knowledge and attitudes into practice to perform activities in an effective, efficient and safe manner.
— Ensuring that the competencies of nuclear industry personnel are developed and maintained.
— Understanding that competencies are built and maintained through a combination of education, training and experience; continuing and refresher training is needed in addition to initial training and performance improvement initiatives.
— Anticipating human resource needs through workforce planning and individual career development (e.g. how to develop succession plans for the organization's future needs and the aspirations of individuals).
— Accountability for the adequacy of training and the performance of personnel.
— Adopting a systematic approach to training as one of the tools to achieve continual performance improvement of tasks by individuals.
— Considering the potentially serious consequences of errors, and why it is particularly important to organize work processes and activities with the defence in depth concept that is central to the nuclear industry (i.e. avoid single points of failure).
— Understanding how employee benefits can provide incentives suitable for retaining personnel and motivating them to contribute to achieving the organization's goals and objectives.
— Awareness of a manager's role to ensure that effective teamwork applies both within the organization, and with suppliers and contractors.
— Effective management of human performance leading to significantly lower rates and consequences of undesirable events.
— Understanding that human error is a normal and expected aspect of human performance but also that accidents result from a combination of factors that are often beyond the control of an individual.
— Focusing on improving both individual and overall organizational performance.
— Organizational culture as a key aspect to achieving good performance.

3.5. Organizational behaviour — The theory and concept of organizational behaviour in the context of the nuclear and radiological sectors and in particular its potential impact on safety, security and performance. This includes consideration of the interaction between individuals and work groups within the organizational structure and setting. This curriculum topic includes issues relating to interdependencies of organizational behaviour along with other aspects of management (e.g. influence of stakeholders, leadership, organizational culture). The suggested teaching topics for this curriculum topic include the following:

— Organizational behaviour dynamics at various levels of an organization;
— Human and organizational factors and their potential role in nuclear safety;
— The potential effects of organizational change on nuclear safety and security;
— Interaction between individuals, technology and organizations;
— Role of a systemic perspective in severe accident management strategies;
— Managing the unexpected through an organizational defence in depth approach;
— The interaction of individuals, technology and organizations in the management of safety;
— Organizational and safety culture and their influence on the management of safety;
— Leadership and management for safety [7] and security;
— The logic of the chain of command and decision making processes in both regulatory bodies and operating licensed nuclear facility organizations during an emergency;
— Potential influence of stakeholders on the basic assumptions, behaviours, competencies and conditions that can affect safety;
— Decision making in team or group settings; how control oriented versus consensus oriented decision making can affect the quality of decisions when risk informed decision making is required, especially for risk-significant operator decisions.

3.6. Financial management and cost control in nuclear or radiological organizations — The financial aspects and related risks associated with nuclear operations or projects and the importance of cost control in the effective management of budgets, scheduling and resources. The suggested teaching topics for this curriculum topic include the following:

— Basic principles of finance, investment, accounting and cost accounting;
— Financial and cost control of key functions to ensure efficient, sustainable and safe projects or business operations (e.g. financial statements, operating expenses versus capital costs, revenue, funding, budgets, tracking expenditures and overheads and levelled life cycle discounted costs, operating costs, fixed costs, variable costs, margins);
— Financial modelling and financial control and system in organizations (e.g. appropriate baseline cost and schedule, work breakdown structure, responsibility assignment matrix, earned value management systems);
— Cost analyses and financial risk analyses (e.g. financial risk involved over the life cycle of nuclear facilities, and related notions such as premium, guarantees, drivers and boost factors, mitigating factors);
— Activity based costing, cost estimating and the time value of money.

3.7. Information and records management in nuclear or radiological organizations — The requirement to understand processes, applications, roles, responsibilities and challenges involved in information and records management in the nuclear and radiological sectors. The suggested teaching topics for this curriculum topic include the following:

— Management information systems (e.g. organizational practice; information types such as contracts, engineering documents and reports on quality assurance; costs associated with information management);
— Documentation management systems (e.g. workflow process, compliance with plant configuration, electronic diagrams, relevant codes and regulations);
— Records management systems (e.g. documentation maintenance, scanning paper copies, intelligent features of scanning of drawings);
— Configuration management systems and operating experience;
— Information and communication technology services, systems, support for cybersecurity (e.g. software selection and its quality assurance, system specification, IT event preparedness);
— Organizational knowledge portals, digital repositories, enterprise application software systems, enterprise resource planning systems.

3.8. Training and human performance management in nuclear or radiological organizations — The training and human performance management aspects, including ensuring that individuals have the competencies needed to perform their assigned tasks, organizing work effectively, and monitoring and continually improving performance. This includes knowledge about the basic principles and tools for excellence in human performance and how those tools should be effectively integrated into all ongoing processes and programmes at a facility to ensure the desired results. The consideration may include performance improvement models, human performance improvement frameworks, nuclear facility personnel training, manager obligations and responsibilities, and the systematic approach to training as a management tool. The impact of any individual's job performance should be considered in thinking about the relevance and importance of human performance for the facility's operation, safety and security. The importance of training on cultural awareness and cross cultural communication should be acknowledged. The suggested teaching topics for this curriculum topic include the following:

— The systematic approach to training process conversant with each of five phases, especially with the first (task analysis) and the last one (evaluation).
— Understanding that the most effective way to learn a new skill or behaviour is to apply knowledge in the workplace and practice skills in real life situations or on a training simulator.
— Human performance management methodology as the basis for a performance continual improvement process. This implies a strong participation of managers in training activities (lectures, classroom observation) and competency field assessments to create a short link between training and actual activity performance.
— Human performance management as a continuous process with event analyses, operating experience, corrective action programmes and self-assessments all providing information which can identify performance improvement opportunities as well as the content of both initial and, especially, continuing training.
— Coaching and mentoring, pre- and post-job reviews, job rotation, task team composition, on-the-job training, refresher training, certification training, use of e-learning.
— Formal qualification training to meet licensing and individual certification requirements.
— Importance of management coaching and mentoring.

3.9. Performance monitoring and organizational improvement — The current performance of an organization, detecting any subtle decline in performance, and looking at all opportunities for improvement by means of self-assessment, performance monitoring, external assessment, independent oversight. It focuses on the ongoing assurance that the management systems at nuclear organizations are effective at ensuring that licensed facilities remain demonstrably within their licence conditions and operate in the safest, most reliable and most cost effective manner possible. The suggested teaching topics for this curriculum topic include the following:

— Periodic reviews of management systems, processes and procedures;
— Plant system health monitoring and surveillance system (plant, systems and equipment);
— Rationale for sound performance indicators (key performance indicators; job, individual, organizational levels);
— Principles and application of human performance indicators and essential notions;
— Documentation and data (e.g. process, self-assessment, review and audits);
— Personnel, training and communication;
— Logistics and support services;
— Periodic safety reviews, peer reviews, incident and event reporting, deficiency reports and corrective action processes;
— Error prevention tools and techniques;
— Relation to quality assurance programmes.

3.10. Nuclear quality assurance programmes — The principles and approaches of quality management systems (i.e. quality assurance, quality control programmes) and their requirements, adoption and implementation as an essential part of an effective management system. Consideration is required as to how they should be applied to all activities affecting the processes and services important to the safety, reliability, performance and security of the facility. This competence also addresses concepts and approaches to the successful implementation of quality assurance systems including planned and systematic actions that provide adequate confidence that the specified requirements are satisfied. The suggested teaching topics for this curriculum topic include the following:

— Quality assurance programmes that include the planned and systematic actions necessary to provide adequate confidence that specified requirements are satisfied;
— Understanding that the implementation of a quality assurance programme involves managers, workers tasks and those responsible for verification and assessment of the effectiveness of the programme, it is not the sole domain of a single group and management has the responsibility to ensure that the programme functions properly; establishing and cultivating principles that integrate practices in daily work activities;
— How to integrate quality assurance into operating experience management;
— International best practices for quality assurance (e.g. International Organization for Standardization — ISO 9001, total quality management programmes);
— What differentiates nuclear quality assurance programmes, especially with respect to design, equipment qualification, and replacement of safety related equipment (e.g. safety requirements including environmental qualification, manufacturing and material control requirements, etc.).

3.11. Procurement and supplier management in nuclear or radiological organizations — The management of the procurement process and the relationships with the suppliers to the nuclear organizations, which have a direct impact on quality assurance at the nuclear facility. All items and services procured must have specified quality requirements to ensure that they do not have an adverse impact on the safety or the operation of the nuclear facility. Due consideration will be required to comply with local procurement procedures. The suggested teaching topics for this curriculum topic include the following:

— Procurement process (identification of needs, technical and quality programme, selection taking into account the short and long term considerations for safety).
— Bid and pre-bid process including soliciting and various in principle agreements.
— All the requirements of the quality assurance programme applied to suppliers. This includes supply of equipment as well as services. Quality requirements must be specified in all contracts including spare part supply, outage subcontractor work and housekeeping services.
— Implementation of quality assurance requirements on suppliers should involve verification of accreditation of supplier personnel coming to work on-site from managers to workers.
— Long term quality partnership between those responsible for verification and assessment of the effectiveness of the programme and suppliers rather than being only driven by cost performance.

3.12. Nuclear safety management and risk informed decision making — The ongoing consideration of the management of safety in the context of management systems. Considerations include proper operating conditions, prevention of accidents, mitigation of accident consequences in order to protect workers, the public and the environment from radiation hazards. There are many sources, locations and hazards of radiation. Knowledge is required for the safe control of radiation hazards in nuclear installations, radioactive waste management and in the transport of radioactive material. This area also addresses management roles and responsibilities to ensure that effective decision and work processes are in place and that adequate organizational resources and accountabilities are established in a manner that specifically addresses safety concerns. These include management of risks under normal circumstances and the consequences of incidents and events that may lead to possible radiological release. The suggested teaching topics for this curriculum topic include the following:

— Sources, locations and types of radiation, and the knowledge required for the safe control of radiation hazards in nuclear installations, radioactive waste management and for the transport of radioactive material.
— Management's role and the responsibilities to ensure that effective decision making and work processes are in place, and adequate organizational resources and accountabilities are established that specifically address safety concerns. Management of risks under normal circumstances, risks as a consequence of incidents, risks as a possible direct consequence of a loss of control over a nuclear reactor core and risks of accident progression.
— The integrated risk informed decision making framework as a systematic process aimed at the integration of the major considerations influencing nuclear power plant safety, with the main goal to ensure that any

decision affecting nuclear safety is optimized without unduly limiting the conduct of operation of the nuclear power plant (including ensuring defence in depth; taking into account research and development and state of the art methodologies; root cause analysis; common cause failure; fail-safe design of systems, structure and components; use of operating experience; safety related processes for components; decision making in normal and abnormal situations; team decision making; regulator's decision making process and its effect on operator safety).

— The range of hazards that pose risks to a range of people in a range of situations at nuclear and radiological facilities. Any decision making process should be clear on how the balancing of different risks is achieved, bearing in mind that measures to reduce one risk may raise others.
— A fundamental aspect of integrated risk informed decision making is that the consequences of decisions affecting safety have to be monitored and feedback needs to be provided on their effectiveness.
— Knowledge and understanding of the methods and basis of the plant safety analyses upon which the design was licensed is needed, and this knowledge, examples listed in the following, should be maintained and relied on for decisions about safety (e.g. how a given accident is prevented or mitigated):
  • Probabilistic risk assessment (including the risk related concept ALARA and project pre-mortem analysis);
  • Deterministic safety assessment (including redundancy and safety margin, engineering and organizational good practices, integration of safety and security);
  • Design basis accidents analyses;
  • Probabilistic safety assessment.

3.13. Nuclear incident management, emergency planning and response — The emergency preparedness and management of nuclear events and the associated emergency response in case of an accident. Integration of safety, security and emergency preparedness programmes provides the optimum protection for public health and safety. The suggested teaching topics for this curriculum topic include the following:

— Intertwined organizations implemented at the facility level, then at the regional authority level, the national authority level and the international level, and the role and responsibility of each stakeholder.
— How nuclear operating organizations declare any event to propose a scaling of the event on the International Nuclear and Radiological Event Scale, to mitigate any associated consequences and to integrate the corresponding lessons into the appropriate operating procedure and training.
— The necessity for the national safety authorities to assess the capabilities of the nuclear facility operator to protect the public by requiring the performance of a full-scale exercise that includes the participation of government agencies.
— How reported events are often routine in nature and do not require activation of an incident response programme; these events are nevertheless analysed and used as lessons to feed a general continuous improvement plan at the plant level and are also shared worldwide.
— Associated communication needs and human reactions during events in general and in discrete events (e.g. case studies of past and recent accident and incidents).
— Emerging requirements for severe accident management (post-Fukushima Daiichi nuclear accident).

3.14. Operating experience feedback and corrective action processes — The managed processes to identify the root causes of past events and prevent the recurrence of similar events. Operating experience is a valuable source of information for learning about and improving the safety, reliability and security of nuclear installations. It focuses on detecting and recording deviations from normal performance by systems and by personnel, especially those which could be precursors of events. It is essential to collect such information in a systematic way for events occurring at nuclear installations during commissioning, operation, maintenance and decommissioning. Corrective action processes are to develop the actions needed to prevent the recurrence of similar events and track their closure and their effectiveness with the aim of continuous improvement. Consideration should also be given to the role of international organizations such as the IAEA, the (OECD Nuclear Energy Agency), the World Association of

Nuclear Operators (WANO) and Institute of Nuclear Power Operations that provide standards and guidelines in this regard. The suggested teaching topics for this curriculum topic include the following:

— Regulatory requirements and the standards and guidelines laid down by international organizations such as the IAEA, OECD/NEA and WANO for any operating experience programme;
— Understanding that, if operating experience is essential to continuous improvement, the field implementation (procedure, training, configuration management) of the corresponding lessons and recommendations is paramount to an effective process;
— Organizational and management elements for an operating experience system, understanding its critical elements, benefits and strategic use (in relation with safety margin, ageing management, business) and potential shortcomings;
— Rationale and international necessity for reporting, data sets for constant feedback loops and possibilities for reviews (e.g. Joint IAEA–OECD/NEA International Reporting System for Operating Experience, IAEA Operational Safety Review Team and Peer Review of Operational Safety Performance Experience missions, WANO peer reviews) and international and regional institutional best practices (e.g. Joint Research Centre's operating experience feedback clearinghouse);
— How to implement efficient operating experience systems including main barriers to the implementation of the operating experience system;
— Purpose and existing practices and processes in terms of operating experience feedback programmes (different approaches and procedures) in nuclear facilities, typical staffing requirements and typical supporting infrastructure (including IT, software);
— Typical recurrent weaknesses identified in establishing successful operating experience (e.g. operational, procedure compliance, human factors, manager engagement, ownership, complacency, process related).

3.15. Nuclear security programme management — The management practices in place to ensure that a nuclear site's security programme is implemented correctly through individual security responsibilities, regulatory compliance and event reporting. The risks to any particular site or their associated facilities will cover all perceived security threats from theft, sabotage, unauthorized access, illegal transfer or other malicious acts involving nuclear material, other radioactive substances, through to protestor disruption, cyberattacks and terrorism. The suggested teaching topics for this curriculum topic include the following:

— Understanding that security is risk based and designed to ensure that appropriate and proportionate controls are implemented and maintained to ensure a safe and secure environment [24];
— Understanding how to balance the needs of safety and security to ensure a safe and secure environment;
— Risk profile that covers all perceived security threats to the business from normal crime and malicious behaviour through to protestor disruption, cyberattacks and terrorism;
— Actual cases of cyber related issues at nuclear power plants (e.g. Slammer worm at Davis Besse).

3.16. Nuclear safety culture — The influences of different factors such as organizational cultural values and norms, individually shared beliefs and perceptions and their impact on the organizational safety and performance. Methods to establish a positive organizational culture should be addressed and included in the training of new employees and the proactive fostering of safety awareness in all employees. A key component of the organizational culture will be the safety culture that requires continuous engagement and dialogue. The suggested teaching topics for this curriculum topic include the following:

— Factors influencing organizational culture and safety culture (e.g. organizational psychology, sociology, anthropology).
— Meaning and link from organizational culture to safety culture.
— Dimensions and characteristics of safety culture (from the organizational to the individual level).
— Influence of a management system on safety culture (e.g. ensuring a common understanding of the key aspects and significance of safety culture in safety performance).
— Dialogue between stakeholders to share information, ideas and knowledge.

— Dialogue enabling the licensee and the regulator to have open discussions with respect to each other's roles. Dialogue supports a more creative and constructive way to find solutions for continuous safety improvements.
— Continuous improvement in safety culture; various levels and definitions of organizational culture and subcultures. Connect these to cultural bias and differences in mental models.
— Different traits in organizational culture and resulting practices and approaches and interactions.
— Ways to understand and address various situations (e.g. lack of collaboration, conflict, disagreement).
— Cross-cultural communication challenges in multicultural environments, international projects.
— Questioning attitude, encouragement of root cause, focused thinking.
— Understanding that a leader should foster a culture for safety in the organization.

3.17. Nuclear events and lessons learned — The key international lessons learned from previous major historical nuclear accidents and their influence on current designs and operational procedures. Awareness and a cursory review of previous major accidents should reinforce the importance of nuclear safety, security and the understanding of hazards and consequences. Specific case studies are encouraged (e.g. the Fukushima Daiichi nuclear accident, the Chernobyl accident, the Three Mile Island accident, and other events). The suggested teaching topics for this curriculum topic include the following:

— Case studies of the causes of all major accidents (the Fukushima Daiichi nuclear accident, the Chernobyl accident and the Three Mile Island accident), near misses and serious incidents (e.g. Davis Besse, Tokai-Mura) and other specific instances;
— Root cause characteristics and precursors of events, and implications on other various nuclear facilities and lessons learned, such as regulatory lessons and operational lessons;
— Regulatory requirements and the standards and guidelines laid down by international organizations such as the IAEA, OECD/NEA and WANO for incident reporting;
— The International Nuclear and Radiological Event Scale as a mechanism to facilitate communication and understanding between the technical community, the media and the public on the safety significance of events.

3.18. Nuclear knowledge management — The establishment of a knowledge management programme and culture within an organization that aligns with a national capacity building policy and strategy, as well as an integrated part of the organizational nuclear infrastructure. This should include topics such as developing knowledge and skills to critically appraise the nature of nuclear knowledge and benefits for the safe operation of nuclear facilities, gains in economics and operational performance, facilitating innovations, and ensuring the responsible use of sensitive knowledge. Consideration should be given to the importance of treating knowledge as an asset, the potential benefits of applying knowledge management tools and techniques, approaches and practices to manage nuclear knowledge, specialized nuclear related information resources and appropriate knowledge management methods and tools. The suggested teaching topics for this curriculum topic include the following:

— Key terms, processes and tools for managing knowledge and its relation to organizational cultures (including knowledge acquisition, transfer, generation, storage);
— Origin, formalization and key drivers for knowledge management across the nuclear and radiological sectors;
— Knowledge management specificities of various nuclear facilities and organizations;
— Various interfaces, tools and processes for the retention of knowledge;
— Instances of strategy implemented (existing knowledge management practices);
— Role of managing knowledge for safety and its relation to organizational performance.

4.   CATEGORY 4: LEADERSHIP

The leadership category includes six curriculum topics and is presented below together with suggested teaching topics. Some requirements for this category were derived from the IAEA publications found in Refs [4–7].

4.1. Strategic leadership — The implementation of the overarching policies of a nuclear or radiological organization. These policies may be aligned with corporate or national strategies and managed by the leadership team, incorporating a strong nuclear safety and security culture. A leader should demonstrate the ability and discipline to solve problems, evaluate options, make judgements and implement a plan using skilful reasoning, accurate information, training and experience. A leader should demonstrate leadership for commitment to safety and security. Effective leadership in nuclear or radiological organizations requires a professional attitude and excellent interpersonal skills. The suggested teaching topics for this curriculum topic include the following:

— Strategic vision aligned with corporate goals, organizational strengths, and corporate social responsibility (e.g. attitude towards safety culture);
— Ability to establish goals, strategies, plans and objectives for the organization that are consistent with the organization's safety policy;
— Ability to establish and respect an effective organizational structure that includes clearly defined roles and responsibilities (e.g. departmental, individual or team, and delegation) and accountability;
— Adequate planning and oversight in executing the mission to meet the strategic vision within the constraints of the external and internal environment;
— The role of leadership in establishing expectations for acceptable management style and conduct, clear expectations for management performance, effective processes for management decision making, and approach to managing risks;
— The role of middle and senior managers in mentoring and developing future leaders in the organization;
— The importance of considering human behaviour, personality types and leadership styles, working relationships and approaches to motivation through recognition, rewards and success sharing, encouraging participative decision making;
— The six principles (purpose, values, method, research, partnership, dialogue) developed by the task force on principles for responsible management education under the coordination of the United Nations Global Compact and leading academic institutions, which lay the foundation for the global platform for ensuring future leaders with the skills needed to balance economics and sustainability;
— Accepting personal accountability in relation to safety on the part of all individuals in the organization and establishing that decisions taken at all levels take account of the priorities and accountabilities for safety;
— Considering case study examples of strategic leadership from the nuclear and radiological sectors;
— Understanding how the Member State's working conditions affect the style and substance around the leadership; understanding the challenge around leadership in the unique nuclear culture environment;

4.2. Ethics and values of a high standard — The principles and behaviours that underpin the decisions, strategies and values embodied in nuclear leadership. Leaders must establish and maintain a strong nuclear safety and security culture in developing good leadership at all levels for their organization. Leaders must foster an environment that promotes accountability and enhances safety performance and security. The suggested teaching topics for this curriculum topic include the following:

— Understanding that establishing a code of ethics and conduct for any organization in nuclear industry is even more important due to the potential safety, security and proliferation hazards unique to this industry, and the role of leaders in this process should be emphasized;
— Considering current nuclear safeguards, security and safety regimes and how the failure to meet these obligations might affect all stakeholders;
— Maintaining meaningful informative dialogue with stakeholders, irrespective of their knowledge level (including the influence of the media, such as social responsibility of nuclear organizations and individuals);
— Understanding the principles of openness, transparency, and accountability of management;
— Considering case studies of past failures of leadership ethics in nuclear and the lessons arising from these failures;
— Encouraging and supporting individuals in the organization to achieve safety goals and perform their tasks safely.

4.3. Internal communication strategies for leaders in nuclear or radiological organizations — Internal communication should be consistent to ensure that the whole workforce understands their role in the attainment of organizational goals. Leaders must communicate clearly the basis for decisions relevant to safety, security and performance. The suggested teaching topics for this curriculum topic include the following:

— Understanding that nuclear safety and security culture is developed and maintained through communication;
— Understanding senior management is responsible for proactively managing communications and stakeholder relationships;
— Implementing communication strategies (i.e. three-way communication) to demonstrate understanding of different cultural, behavioural, and other biases in communication;
— Using different communication pathways for effective communication;
— Communicating clearly the basis for decisions relevant to safety; enabling safety discussion with real examples;
— Considering specific case studies where leadership communication was important to success;
— Understanding the importance of careful communication when balancing safety, reliability, and economics and production;
— Understanding the importance of promotion of dialogue and honest, open communication;
— Supporting a questioning attitude.

4.4. External communication strategies for leaders in nuclear or radiological organizations — External communication should be clear and transparent and developed with stakeholders, recognizing the prevailing environment affecting the operation of an organization (e.g. socioeconomic, political). The suggested teaching topics for this curriculum topic include the following:

— Understanding that external communication should be clear and transparent and developed with stakeholders, recognizing the prevailing environment affecting the operation of an organization (e.g. socioeconomic, political);
— Understanding the responsibility of proactively managing stakeholder relationships;
— Using a continuous public relations strategy as a key method of communication with stakeholders (including the media's role);
— Different communication pathways (such as social media tools) and their effect on the message transmitted and received in stakeholder dialogue.

4.5. Leading change in nuclear or radiological organizations — A plan that is aligned to the strategic vision. Effective communication is required to ensure that the risks and opportunities relating to change are understood by all stakeholders, and how they may affect the safety and security of the nuclear organization. Examples may include internal reorganization, merger, acquisition or joint venture. The suggested teaching topics for this curriculum topic include the following:

— Leaders should comprehend the risks and opportunities that changes within an organization may bring and the effects that any changes might have on the safety and security of the organization, and develop leadership strategies to manage these altered circumstances;
— Change management policy should give priority to safety and impact of change on safety;
— Observation skills are needed to recognize where changes occur in internal and external stakeholders (e.g. IAEA, OECD/NEA, WANO);
— Gap analysis skills are needed to identify needs and drivers for change;
— The role of communication during organizational change;
— The need for inspections and tools for oversights of change management;
— The need to consider and proactively manage the risk of critical knowledge loss associated with change;
— The need to consider specific case studies of organizational change.

4.6. Leadership to support the safety culture — An environment that promotes accountability and enhances safety performance and security. A leader should demonstrate leadership for safety and a commitment to safety and security. The suggested teaching topics for this curriculum topic include the following:

— Acceptance of personal accountability in relation to safety on the part of all individuals in the organization and establishing that decisions taken at all levels take account of the priorities and accountabilities for safety

— Understanding of current nuclear safeguards, security and safety regimes and how the failure to meet these obligations might affect all stakeholders;

— Encouraging and supporting individuals in the organization to achieve safety goals and perform their tasks safely;

— Understanding of the risks and opportunities that changes within an organization may bring and the effects that any changes might have on the safety and security of the organization, and develop leadership strategies to manage these altered circumstances;

— Understanding the need for a change management policy giving priority to safety and impact of change on safety;

— Understanding the need to consider specific case studies, including crisis management and lessons learned.

Appendix II

ASSIST MISSION SELF-ASSESSMENT TOOL AND FORM TEMPLATE

The INMA preliminary self-assessment tool (Fig. 1) and preliminary programme description form (Fig. 2) provide information on a university's plans and status for the establishment of their INMA-NTM programme. They are required to be completed and submitted to the Nuclear Knowledge Management Section of the IAEA six weeks prior to an assist mission.

INMA preliminary self-assessment tool for the existing master's programme(s) at [name of university]																
Compiled by:		For each existing university course please provide a rating of your staff's capability to teach each curriculum topic, add more programmes as required. Leave blank for no capability, 1 for some capability, 2 for adequate capability and 3 for excellent capability,														
Date:		Name of programme 1					Name of programme 2					Name of programme 3				
INMA International Nuclear Management Academy **IAEA**	INMA requirements R = required curriculum topic A = as appropriate	Course 1	Course 2	Add column as necessary, one column for each course												
Category 1. External environment																
1.1 Energy production, distribution and markets	A															
1.2 International nuclear and radiological organizations	R															
1.3 National nuclear technology policy, planning and politics	A															
1.4 Nuclear standards	R															
1.5 Nuclear and radiological law	A															
1.6 Business law and contract management	R															
1.7 Intellectual property management	A															
1.8 Nuclear and radiological licensing, licensing basis and regulatory processes	R															
1.9 Nuclear security	R								•							
1.10 Nuclear safeguards	A															
1.11 Transport of nuclear goods and materials	A															
Category 2. Technology																
2.1 Nuclear or radiological facility design principles	R															
2.2 Nuclear or radiological facility operational systems	R															
2.3 Nuclear or radiological facility life management	A															
2.4 Nuclear or radiological facility maintenance processes and programmes	R			•												
2.5 Systems engineering for nuclear or radiological facilities	A															
2.6 Nuclear safety principles and analysis	R															
2.7 Radiological safety and protection	R															
2.8 Nuclear reactor physics and reactivity management	A															
2.9 Nuclear fuel cycle technologies	A															
2.10 Radioactive waste management and disposal	R															
2.11 Nuclear or radiological facility decommissioning	R															
2.12 Environmental protection, monitoring and remediation	R															
2.13 Nuclear research and development and innovation management	A															
2.14 Application of nuclear science	A															
2.15 Thermohydraulics	A															
Category 3: Management																
3.1 Nuclear engineering project management	R															
3.2 Management systems in nuclear or radiological organizations	R															
3.3 Management of employee relations in nuclear or radiological organizations	R															
3.4 Organizational human resource management and development	R															
3.5 Organizational behaviour	R															
3.6 Financial management and cost control in nuclear or radiological organizations	R															
3.7 Information and records management in nuclear or radiological organizations	R															
3.8 Training and human performance management in nuclear or radiological organizations	R															
3.9 Performance monitoring and organization improvement	R															
3.10 Nuclear quality assurance programmes	R															
3.11 Procurement and supplier management in nuclear or radiological organizations	R															
3.12 Nuclear safety management and risk informed decision making	R															
3.13 Nuclear incident management, emergency planning and response	R															
3.14 Operating experience feedback and corrective action processes	R															
3.15 Nuclear security programme management	A															
3.16 Nuclear safety culture	R															
3.17 Nuclear events and lessons learned	R															
3.18 Nuclear knowledge management	R															
Category 4: Leadership																
4.1 Strategic leadership	R															
4.2 Ethics and values of a high standard	R															
4.3 Internal communication strategies for leaders in nuclear or radiological organizations	R															
4.4 External communication strategies for leaders in nuclear or radiological organizations	R															
4.5 Leading change in nuclear or radiological organizations	R															
4.6 Leadership to support the safety culture	R															

FIG. 1. INMA preliminary self-assessment tool.

INMA	**INMA Preliminary Programme Description Form**
International Nuclear Management Academy	Please answer all the questions that are applicable to the current status of the programme

1. General information			
Compiled by		Date	
University			
Programme title			
Academic level			
Faculty/school/department			

2. Policy, strategy and vision

i. Describe your university's policy, strategy and vision and to what extent it is aligned with national policy.

Enter response here.

ii. What are the national and international dimensions of your university(s), department(s) and programme?

Enter response here.

iii. Describe any participation in educational and professional networks, forums, seminars and conferences.

Enter response here.

iv. Is there a knowledge management policy, and if so, how is it implemented?

Enter response here.

3. Programme entry requirements

i. Describe the strategy for attracting, selecting and enrolling students and whether there is a strategy to attract the highest calibre students.

Enter response here.

ii. Will scholarships for students be available and if so what are the selection criteria?

Enter response here.

iii. Which qualifications, and in which subjects, will be acceptable for students to enter the programme?

Enter response here.

iv. Will students need work experience before entering the programme?

Enter response here.

v. What is the minimum and maximum number of students that will be able to enter the programme each year?

Enter response here.

vi. What will be the fee for the programme? Please include all components such as entrance fee and tuition fee and explain what they cover.

Enter response here.

vii. Who will typically pay the fee, student or employer?

Enter response here.

viii. Will the fee be set by the Government or the university and how does it compare to other similar programmes at the university?

Enter response here.

ix. Are there any foundation courses that can be taken, if necessary, before entering the nuclear technology management programme?

Enter response here.

4. Programme delivery

i. How will quality staff be attracted to the programme, and what are the policies to retain them?

Enter response here.

ii. Will industry provide any lecturers for the nuclear technology management programme?

Enter response here.

iii. Will the experience of nuclear experts close to retirement or already retired be utilized?

Enter response here.

iv. What is the target student-teacher ratio for lectures, tutorials and practices?

Enter response here.

v. Describe the experimental facilities of the university, or access arrangements, to experimental facilities, simulators and computer research facilities.

Enter response here.

vi. How will you utilize libraries and IAEA publications in the delivery of the programme?

Enter response here.

vii. List some recent relevant peer-reviewed publications authored by the department/university.

Enter response here.

i. Will the nuclear technology management post-graduate programme be part-time or full-time etc.?

Enter response here.

ii. Will the programme be flexible enough to allow students the time to supplement their income through working to support their studies?

Enter response here.

iii. Please provide an outline timetable of the academic week and year. If there is an elective portion of the study, please provide indicative timetables.

Enter response here.

iv. Provide details on how the courses will be delivered (academic lectures, combination of lectures, tutorials, quizzes and industry lectures and will it use supporting material via videos, e-learning etc).

Enter response here.

v. What will be the credit structure for the programme and how many hours of total study are required for one credit?

Enter response here.

vi. List the planned mandatory and elective courses.

Enter response here.

vii. Will students choose their dissertation from a list or will they develop their own project?

Enter response here.

viii. Will the projects be based in industry and/or university?

Enter response here.

ix. How will the nuclear industry needs be reflected in the curricula?

Enter response here.

x. How will soft skills (e.g. communication, team work, etc.) be incorporated into the programme?

Enter response here.

6. Quality control

i. Describe any proposed quality control methodology for the programme.

Enter response here.

ii. Describe any proposed external accreditation.

Enter response here.

iii. How will student feedback be gathered and reported, and how will it affect the development and delivery of the programme?

Enter response here.

7. National and international dimensions

i. What interaction exists with national and international educational or research organizations, bodies/agencies and learned/professional associations?

Enter response here.

ii. Will students be able to undertake any part of their studies at other national or international educational organizations, and if so, provide details?

Enter response here.

iii. Do any of the staff teach at other national or international educational organizations?

Enter response here.

iv. Are there any memorandums of understanding/agreements with other national or international educational organizations related to nuclear technology management?

Enter response here.

v. Will any part of the nuclear technology management programme be taught by visiting staff from other national or international organizations?

Enter response here.

8. Collaboration with industry

i. Describe how industrial collaboration will contribute to the delivery of the programme.

Enter response here.

ii. Will there be an external advisory board for the nuclear technology management programme and if so, what will be its composition, role, input and frequency of meetings?

Enter response here.

iii. Is it planned for industry to provide any internships and/or offer support for theses (diploma) work by providing project placements with associated costs?

Enter response here.

iv. Will industry support students through prizes, awards, scholarships, etc.?

Enter response here.

v. Will the university offer any nuclear technology management short courses to industry for employee continual professional development (CPD)?

Enter response here.

FIG. 2. Preliminary programme description form.

Appendix III

ASSESSMENT MISSION SELF-ASSESSMENT TOOL AND FORM TEMPLATES

The INMA programme self-assessment tool (Fig. 3), programme description form (Fig. 4), course description form (Fig. 5) and courses delivery and assessment forms (Fig. 6) provide information on a university's NTM programme. they are required to be completed and submitted to the Nuclear Knowledge Management Section of the IAEA six weeks prior to an assessment mission.

INMA NTM programme self-assessment tool for the [name of programme] at the [name of university]																		

Compiled by:

Date:

	INMA requirements R = required curriculum topic A = as appropriate	Total learning hours for each curriculum topic	**Core courses**					**Elective courses**					**Elective courses** (offered by third parties)				
			Course 1	Course 2	Add columns as necessary, one column for each course			Add or delete columns as necessary, one column for each course					IAEA NKM or NEM School	WNU Summer Institute	IAEA Nuclear Law Institute	OCED/NEA-Montpellier University Int'l School of Nuclear Law	Other (e.g. MIT INLEP course)
Learning hours breakdown: (for definitions please see the instructions sheet) % Direct teaching : % Self study : % Practical exercise			00:00:00	00:00:00	00:00:00	00:00:00	00:00:00	00:00:00	00:00:00	00:00:00	00:00:00	00:00:00	00:00:00	00:00:00	00:00:00	00:00:00	00:00:00
Direct teaching delivering mode: In person (I), Online (O), or Both (B)																	
Teaching approach: Theoretical-conceptual (TC), Experiential-applied (EA), or Both (B)																	
Expected number of hours of lectures from industry experts and leaders: specify the number																	
Total learning hours for each course			0	0	0	0	0	0	0	0	0	0					
Category 1. External environment																	
1.1 Energy production, distribution and markets	A	0															
1.2 International nuclear and radiological organizations	R	0															
1.3 National nuclear technology policy, planning and politics	A	0															
1.4 Nuclear standards	R	0															
1.5 Nuclear and radiological law	A	0															
1.6 Business law and contract management	R	0															
1.7 Intellectual property management	A	0															
1.8 Nuclear and radiological licensing, licensing basis and regulatory processes	R	0															
1.9 Nuclear security	R	0															
1.10 Nuclear safeguards	A	0															
1.11 Transport of nuclear goods and materials	A	0															
Category 2. Technology																	
2.1 Nuclear or radiological facility design principles	R	0															
2.2 Nuclear or radiological facility operational systems	R	0															
2.3 Nuclear or radiological facility life management	A	0															
2.4 Nuclear or radiological facility maintenance processes and programmes	R	0															
2.5 Systems engineering for nuclear or radiological facilities	A	0															
2.6 Nuclear safety principles and analysis	R	0															
2.7 Radiological safety and protection	R	0															
2.8 Nuclear reactor physics and reactivity management	A	0															
2.9 Nuclear fuel cycle technologies	A	0															
2.10 Radioactive waste management and disposal	R	0															
2.11 Nuclear or radiological facility decommissioning	R	0															
2.12 Environmental protection, monitoring and remediation	R	0															
2.13 Nuclear research and development and innovation management	A	0															
2.14 Application of nuclear science	A	0															
2.15 Thermohydraulics	A	0															
Category 3: Management																	
3.1 Nuclear engineering project management	R	0															
3.2 Management systems in nuclear or radiological organizations	R	0															
3.3 Management of employee relations in nuclear or radiological organizations	R	0															
3.4 Organizational human resource management and development	R	0															
3.5 Organizational behaviour	R	0															
3.6 Financial management and cost control in nuclear or radiological organizations	R	0															
3.7 Information and records management in nuclear or radiological organizations	R	0															
3.8 Training and human performance management in nuclear or radiological organizations	R	0															
3.9 Performance monitoring and organization improvement	R	0															
3.10 Nuclear quality assurance programmes	R	0															
3.11 Procurement and supplier management in nuclear or radiological organizations	R	0															
3.12 Nuclear safety management and risk informed decision making	R	0															
3.13 Nuclear incident management, emergency planning and response	R	0															
3.14 Operating experience feedback and corrective action processes	R	0															
3.15 Nuclear security programme management	A	0															
3.16 Nuclear safety culture	R	0															
3.17 Nuclear events and lessons learned	R	0															
3.18 Nuclear knowledge management	R	0															
Category 4: Leadership																	
4.1 Strategic leadership	R	0															
4.2 Ethics and values of a high standard	R	0															
4.3 Internal communication strategies for leaders in nuclear or radiological organizations	R	0															
4.4 External communication strategies for leaders in nuclear or radiological organizations	R	0															
4.5 Leading change in nuclear or radiological organizations	R	0															
4.6 Leadership to support the safety culture	R	0															
External environment total learning hours		0															
Technology total learning hours		0															
Management total learning hours		0															
Leadership total learning hours		0															
Programme total learning hours		0															

FIG. 3. INMA programme self-assessment tool.

| INMA |
| International Nuclear Management Academy |

INMA Programme Description Form

Please answer all the questions that are applicable to the current status of the programme

1. General information			
Compiled by		Date	
University			
Programme title			
Academic level			
Faculty/school/department			

2. Policy, strategy and vision

i.   Describe your university's policy, strategy and vision and to what extent it is aligned with national policy.

Enter response here.

ii. What are the national and international dimensions of your university(s), department(s) and programme?

Enter response here.

iii. Describe any participation in educational and professional networks, forums, seminars and conferences.

Enter response here.

iv. Is there a knowledge management policy, and if so, how is it implemented?

Enter response here.

3. Programme entry requirements

i. Describe the strategy for attracting, selecting and enrolling students and whether there is a strategy to attract the highest calibre students.

Enter response here.

ii. Are scholarships for students available and if so what are the selection criteria?

Enter response here.

iii. Which qualifications, and in which subjects, are acceptable for students to enter the programme?

Enter response here.

iv. Do students need work experience before entering the programme?

Enter response here.

v. What is the minimum and maximum number of students that can enter the programme each year?

Enter response here.

vi. What is the fee for the programme? Please include all components such as entrance fee and tuition fee and explain what they cover.

Enter response here.

vii. Who typically pays the fee, student or employer?

Enter response here.

viii. Is the fee set by the Government or the university and how does it compare to other similar programmes at the university?

Enter response here.

ix. Are there any foundation courses that can be taken, if necessary, before entering the nuclear technology management programme?

Enter response here.

4. Programme delivery

i. How are quality staff attracted to the programme, and what are the policies to retain them?

Enter response here.

ii. What are the demographics (age distribution, gender, etc.) and academic experience of the nuclear technology management programme staff?

Enter response here.

iii. Has a risk assessment for the loss of experienced teaching staff and their knowledge been undertaken?

Enter response here.

iv. Does industry provide any lecturers for the nuclear technology management programme?

Enter response here.

v. How is the experience of nuclear experts close to retirement or already retired utilized?

Enter response here.

vi. What is the student-teacher ratio for lectures, tutorials and practices?

Enter response here.

vii. Describe the experimental facilities of the university, or access arrangements, to experimental facilities, simulators and computer research facilities.

Enter response here.

viii. How do you utilize libraries and IAEA publications in the delivery of the programme?

Enter response here.

ix. List some recent relevant peer-reviewed publications authored by the department/university.

Enter response here.

5. Curricula

i.   Is the nuclear technology management post-graduate programme part-time or full-time etc.?

Enter response here.

ii.   Is the programme flexible enough to allow students the time to supplement their income through working to support their studies?

Enter response here.

iii.   Please provide a timetable of the academic week and year. If there is an elective portion of the study, please provide indicative timetables.

Enter response here.

iv.   Provide details on how the courses are delivered (academic lectures, combination of lectures, tutorials, quizzes and industry lectures and does it use supporting material via videos, e-learning etc).

Enter response here.

v.   What is the credit structure for the programme and how many hours of total study are required for one credit?

Enter response here.

vi.   List the mandatory and elective courses and provide statistics on the number of students choosing each elective course.

Enter response here.

vii.   Provide examples of the assessments and assignments set and a list of dissertation titles.

Enter response here.

viii.   Do students choose their dissertation from a list or can they develop their own project?

Enter response here.

ix.   Are the projects based in industry and/or university?

Enter response here.

x.   How are the nuclear industry needs reflected in the curricula?

Enter response here.

xi.   How are soft skills (e.g. communication, team work, etc.) incorporated into the programme?

Enter response here.

6. Quality control

i. Describe any quality control methodology for the programme.

Enter response here.

ii. Describe any external accreditation and the type of information provided by the university for the accreditation.

Enter response here.

iii. Are the recommendations offered by the accrediting organization mandatory for the university?

Enter response here.

iv. How is student feedback gathered and reported, and how does it affect the development and delivery of the programme?

Enter response here.

7. National and international dimensions

i.   What interaction exists with national and international educational or research organizations, bodies/agencies and learned/professional associations?

Enter response here.

ii. Are students able to undertake any part of their studies at other national or international educational organizations, and if so, provide details?

Enter response here.

iii. Do any of the staff teach at other national or international educational organizations?

Enter response here.

iv. Are there any memorandums of understanding/agreements with other national or international educational organizations related to nuclear technology management?

Enter response here.

v.   Is any part of the nuclear technology management programme taught by visiting staff from other national or international organizations?

Enter response here.

8. Collaboration with industry

i. Describe the collaboration with industry and how it contributes to the delivery of the programme.

Enter response here.

ii. Is there an external advisory board for the nuclear technology management programme and if so, what is its composition, role, input and frequency of meetings?

Enter response here.

iii. What is the type of feedback sought from industry on the programme?

Enter response here.

iv. Does industry provide any internships and/or offer support for theses (diploma) work by providing project placements with associated costs?

Enter response here.

v. Does industry support students through prizes, awards, scholarships, etc.?

Enter response here.

vi. Does the university offer any short courses to industry for employee continual professional development (CPD)?

Enter response here.

9. Outcomes

i. What are the student failure and graduation rates for the courses?

Enter response here.

ii. Is there a strategy in place to track and maintain interaction with alumni?

Enter response here.

iii. What are the students' professional destinations after graduating?

Enter response here.

iv. How many graduates enter, or remain, in the nuclear profession after completing the programme?

Enter response here.

FIG. 4. INMA programme description form.

INMA Course Description Form

1. General information			
Compiled by		Date	
University			
Programme title			
Faculty/school/department			
Name of course			
Total direct teaching hours for the course			
Total learning hours for the course			
Number of credits for the course			
Name of course director			
Names of the academic lecturer(s)			
Names of the industrial lecturer(s)			

2. Aims of the course

Enter response here.

3. Description of the course and the course subjects

Enter response here.

4. General learning outcomes of the course

Knowledge of a subject (*Bloom's taxonomy - knowledge and comprehension*)
Example phrases: remembers previously learned material, understands the meaning of material.

Enter response here.

Demonstration of the application of knowledge (*Bloom's taxonomy - application and analysis*)
Example verbs: apply, carry out, demonstrate, illustrate, prepare, solve, use.

Enter response here.

How and when to implement the knowledge (*Bloom's taxonomy - synthesis and evaluation*).
Example verbs: combine, construct, design, develop, generate, plan, propose, assess, conclude, evaluate, interpret, justify, select, support.

Enter response here.

FIG. 5. INMA course description form.

INMA Courses Delivery and Student Assessment Description Form

1. General information		
Compiled by		Date
University		
Programme title		

2. Delivery methodology

Please enter the number of hours for each type of delivery method	Direct teaching				Practical exercise		Self study					Total hours for the course
	Lectures	Seminars	Academic tutorials	Personal tutorials	Laboratory work	Computer exercises	Individual projects	Group projects	Field study	Self-directed study and revision	Other	
Course 1												1
Course 2												2
Course 3												3
Etc.												0
												0

Please enter the number of hours for each type of delivery method	Direct teaching				Practical exercise		Self study					Total hours for the course
	Lectures	Seminars	Academic tutorials	Personal tutorials	Laboratory work	Computer exercises	Individual projects	Group projects	Field study	Self-directed study and revision	Other	
												0
												0
												0
												0
Total Hours	0	0	0	0			0	0	0	0	0	0

3. Assessment											
Please enter the number of hours for each type of student assessment	Examination	In class test	Online quiz	Assessed seminar work	Written report	Oral presentation	Poster	Practical demonstration	Attendance	Other	Total hours for the assessment
Course 1											1
Course 2											2
Course 3											3
Etc.											0
											0
											0
											0
											0
											0
Total Hours	0	0	0	0	0	0	0	0	0	0	0

FIG. 6. INMA courses delivery and assessment forms.

Appendix IV

MISSION AGENDA TEMPLATE

The typical topics and sequence that would be included in a mission agenda are given in this appendix:
Day 1:

— Welcome and opening remarks;
— Overview of INMA (IAEA Scientific Secretary);
— Presentations by INMA stakeholders (e.g. industry representatives);
— Methodology of the mission and assessment process (IAEA Scientific Secretary);
— Objectives, expected outcomes and draft agenda of this mission (IAEA Scientific Secretary);
— Presentations by the mission team members on their universities' INMA-NTM or NTM programme.

Days 2–4 (the detailed programme — discussions and comprehensive review of the information package course by course):

— Discussions should be based on, and provide further explanation where necessary on, the previously submitted information package;
— An overview of the host university's NTM programme, including the structure of the programme and choice of programmatic theme, if any;
— The host university's strategy for the implementation of their NTM programme;
— Details on each programme course/module (including mandatory and elective courses/modules) are discussed referring to the previously submitted course description forms (e.g. subject topics/content and estimated learning hours, curriculum topics, delivery method, number of credits, duration, staff);
— Details on student projects, internship, thesis, diploma (e.g. student research work, terminal or non-terminal degree);
— Admission and evaluation of prospective students;
— The e-learning methodology, if applicable;
— The pedagogical approaches used in the programme;
— Collaboration mechanisms with industry and national and international partners;
— Student presentations;
— Industry demand and interest in the NTM programme;
— Review of the information package (the self-assessment tools, programme description form and course description forms);
— Visit to possible facilities (e.g. nuclear power plant simulator laboratory);

Day 5:

— Mission team to draft report outline, to discuss and summarize their preliminary observations;
— Q&A.

Appendix V

GUIDANCE FOR INMA MISSION PREPARATION

Suggested guidance for the preparation of an INMA mission is provided below:

(a) The focus of the mission is not on a detailed review of the curriculum and content of each individual course, rather the objective is to explain the logistics of each course and how it fits into the overall programme (visiting team members are familiar with the concepts being taught). The detailed curriculum and content should be presented in the information package.

(b) The focus should be on why the course has been developed and why the topics taught have been chosen; consider if adequate stakeholder engagement and input was undertaken (or was the programme just based on the strengths of the department, faculty or university).

(c) It is not necessary to make detailed presentations on the content of each of the courses. Presentations on selected courses can be discussed with the host university after the information package has been reviewed .

(d) The information package should be submitted six weeks ahead of the mission. There should be a pre-mission teleconference discussion between the IAEA Scientific Secretary and the mission team with the university counterpart as to which courses have detailed presentations and plan the agenda.

(e) When a presentation on a course is requested, a simple list of the topics included in the course is sufficient as the course description forms are included in the information package (i.e. one or two sample slides are useful to illustrate the lecture content.) Any particular aspect of a course that is of interest to the visiting team can be elaborated.

(f) The pre-requisites for students to enter the programme should be clearly stated in the information package and presented during the mission (e.g. the bachelor's degrees that are acceptable). The approach as to how the various competencies of the students on entry are made equivalent (e.g. make-up courses, on-boarding using the Nuclear Energy Management School as an integrated part of the programme). It should be clear whether students need work experience before entering the programme. Requirements and rationale should be clear in the information package and presented during the mission.

(g) The minimum and maximum number of students that can enter the programme each year should be stated, as should the average number of students expected and number students that have successfully completed the programme each year.

(h) It should be clear how the courses are to be delivered, whether they are just academic lectures or a combination of lectures, tutorials, quizzes, industry lectures, and whether the course uses supporting material through media such as videos or e-learning.

(i) The credit structure for the degree programme should be clearly explained including how many hours of total study are required for one credit in the local university system, as this varies from country to country. The mission team and other INMA members need to understand the established credit system.

(j) It should be clear for each course how many hours correspond to lectures and how many correspond to self-study. Is a consistent ratio of lectures to overall learning hours used for each course? The spreadsheet templates in the information package help to show this.

(k) Basic information about the university's academic year should be explained and a timetable of key annual dates should be provided. If there is an elective portion of the study, an indicative or common timetable for the years of the programme should be provided. If part-time studies are permitted, any limitations on the programme duration should be explained.

(l) The typical academic week should be explained and representative timetables should be provided. If there is an elective portion of the study, provide an indicative or common timetable for the weeks of the programme.

(m) If the programme has mandatory and elective courses, these must be classified and distinguished clearly. The mission will assess how the core elements of the programme meet the curriculum topic requirements.

(n) Provide statistics on the number of students choosing each of the elective courses, if possible.

(o) Provide examples of the assessments and assignments set for each course, if possible.

(p) Provide a list of past student projects or dissertation titles.

(q) Explain how the dissertation topics are chosen; clarify if the students choose from a list or are they developing their own project ideas (e.g. clarify whether the student projects are based on industry or university research topics).

(r) Summarize the fee structure for the programme. Include all components if there are multiple components such as entrance fees and tuition fees and explain what the various fees cover.

(s) Clarify who typically pays the student fee (i.e. the student, a government subsidy, the employer or some combination thereof). Clarify how the fee compares to other similar programmes at the university and whether the fee is set by the government or university or faculty.

(t) If possible, provide data or examples of the student employment rates and destinations after graduating.

(u) Regarding the mission agenda, it should be based on the template agenda provided in the publication. Some key points are listed in the following:

  (i) First item on the agenda should be an introduction of all participants that includes their role in the mission.

  (ii) The mission team members would not normally give presentations on their INMA-NTM programmes unless specifically requested by the host university (e.g. for the benefit of stakeholders that will be present).

  (iii) There should not be a detailed IAEA presentation on the INMA programme unless specifically requested or required (e.g. if senior management or other stakeholders are present).

  (iv) The objectives and process of the INMA missions and endorsement process should be reiterated at the start of the meeting to the team and the local faculty involved in the mission.

An overview presentation on INMA can be of value to explain the global nature, reason for the establishment of INMA and its aims, and to highlight other participating member universities of INMA to present a bigger picture, but this is needed only if senior management are present, for example typically in the opening session.

Appendix VI

ASSIST MISSION REPORT TEMPLATE

The assist mission report template (Fig. 7) has been designed to help the mission team compile their report on completion of their visit to the university.

Assist Mission Report

for the

[name of NTM master's programme]

at the

[name of the university]
[country]

From [day] to [day], [month], [year]

DEPARTMENT OF NUCLEAR ENERGY
DIVISION OF PLANNING, INFORMATION AND KNOWLEDGE MANAGEMENT
NUCLEAR KNOWLEDGE MANAGEMENT SECTION
IAEA 2020

CONTENTS

1. ADMINISTRATIVE INFORMATION

Dates of assist mission: [dd to dd, month, 20yy]

Duty station:

Local counterpart:

INMA assist mission team

Country	Participant/organization	E-mail	Remarks
			Local host
			Scientific Secretary

2. BACKGROUND

The IAEA facilitates the International Nuclear Management Academy (INMA) for nuclear engineering and science universities that provide master's degree programmes focusing on technology management for the nuclear sector. The purpose of INMA is to improve the safety, performance and economics of nuclear technologies by promoting and enabling the availability and accessibility of high quality educational programmes for managers working in the nuclear sector and improving their management competencies.

The IAEA's Nuclear Knowledge Management (NKM) Section and participating universities have identified and defined a set of curriculum topics to help universities design INMA compliant nuclear technology management (NTM) master's programmes. The IAEA offers support and assistance to universities to implement NTM programmes and coordinates with the INMA participating universities on the development and maintenance of the requirements for INMA-NTM programmes by organizing meetings and missions.

An initial assist mission upon request by a university can review its current master's programmes that could contribute to an NTM programme as well as the university's plan to implement an NTM programme. The mission will help the university to fully understand the requirements, to share the experiences of other INMA members and to provide guidance on programme design.

The [name of university] has requested the IAEA to conduct an INMA initial assist mission. The mission was organized from [dd to dd, month, 20yy] to assess the feasibility of implementing a master's programme to meet the INMA requirements and to establish organizational support and a network between the university and the stakeholders. The mission provided support to the university and their stakeholders, to understand the INMA practices and facilitate the development of an INMA-NTM programmes.

3. OBJECTIVES

The objectives of this INMA initial assist mission were to:

- Present the INMA collaboration framework and the detailed NTM requirements to the [name of university] and interested stakeholders.
- Obtain a high level understanding and inventory of the university's existing programmes related to NTM, including their implementation and approach and applicability to their proposed future INMA master's degree programme in NTM;
- Assist the university in an assessment of the gaps between their existing programmes and the INMA common requirements for NTM programmes, and discuss possible improvements;
- Provide input and feedback on the design of the programme to strengthen the university's future programme in NTM as well as to the IAEA on INMA;
- Share best practices and the experiences of other universities participating in INMA with respect to various aspects and elements related to the implementation of INMA programmes.

4. DETAILS OF THE PROPOSED UNIVERSITY NTM PROGRAMME

4.1. Highlights of the discussions with the university's representatives and nuclear industry stakeholders

[normally compiled in chronological order]

4.2. Status of the development of a nuclear technology management master's level programme at the [name of university]

5. EXISTING UNIVERSITY COURSES THAT MAY POSSIBLY BE USED WITHIN THE NTM PROGRAMME

6. EQUIPMENT, FACILITIES AND HUMAN RESOURCES

7. RECOMMENDATIONS TO THE UNIVERSITY ON HOW TO FULLY DEVELOP ITS INMA-NTM PROGRAMME

8. APPENDICES

8.1. Appendix 1: Meeting agenda.

8.2. Appendix II: List of participants (with photographs if possible)

8.3. Appendix III: Preliminary information package documents

Below is a list of questions that should ideally be answered within the report:

- What is the format of the programme?
- What are the entrance requirements?
- Which bachelor's degree subjects are accepted as NTM entry requirements?
- How are the different backgrounds and capabilities of the students expected to be managed within the NTM programme?
- Is it a part-time or full-time programme?
- How many courses are required for a full master's degree?
- Are they all the same number of hours?
- How many study hours are required for a full master's degree?
- How many hours are required for the project?
- Is it one year or two years?
- How many hours learning equals one credit?
- Is the programme semester based?
- Is a weekly and yearly timetable (student schedule) available and when does the academic year start?
- When do exams take place?
- Are exams compulsory for each course?
- Is distance learning available?
- If so, how much of the programme?
- What is the funding for the programme?
- What will be the fee for the master's programme?
- Who pays the fees, is it the students themselves or sponsoring companies?
- Are bursaries available?
- What is the forecast for the number of students each year?
- What is the minimum number of students to sustain the programme?
- What is the expected ratio of university-stakeholder lecturers?
- What is the target date for the first intake of students?
- Will the programme be available for international students?
- Will the programme be delivered in English?

FIG. 7. INMA assist mission report template.

Appendix VII

ASSESSMENT MISSION REPORT TEMPLATE

The assessment mission report template (Fig. 8) has been designed to help the mission team compile their report on completion of their visit to the university.

Assessment Mission Report

on the

[name of NTM master's programme]

at the

[name of the university]
[country]

From [day] to [day], [month], [year]

DEPARTMENT OF NUCLEAR ENERGY
DIVISION OF PLANNING, INFORMATION AND KNOWLEDGE MANAGEMENT
NUCLEAR KNOWLEDGE MANAGEMENT SECTION
IAEA 2020

CONTENTS

EXECUTIVE SUMMARY

This report summarizes the activities and suggested actions of the IAEA coordinated International Nuclear Management Academy (INMA) assessment mission to the [name of university] conducted from [dd to dd, month, 20yy]. The objective of the mission was to review the master's programme in nuclear technology management (NTM), evaluate its compliance as an INMA-NTM programme, consider the university's overall approach and implementation of their programme and give suggestions for further improvement.

The review was based on the IAEA Nuclear Energy Series publication that defines the requirements for INMA-NTM programmes and the procedure for the INMA endorsement process [add here the relevant reference number within this report for the INMA NE Series document]. The university submitted a complete information package to the IAEA for review in advance of the mission.

The mission team was led by [name of the Scientific Secretary], the IAEA Scientific Secretary, and included senior academics that have been participating in the development of INMA. The team included [list names, organization and country].

The assessment scope included [please describe the scope]. The comprehensive assessment took place over [no. of days] days [dd to dd, month, 20yy] at [location]

The mission team concluded that [summary of findings].

1. ADMINISTRATIVE INFORMATION

Date of mission: [dd to dd, month, 20yy]

Duty station:

Counterpart:

INMA assessment mission team:

Country	Participant/organization	E-mail	Remarks
			Local host
			Scientific Secretary

2. BACKGROUND

2.1 The International Nuclear Management Academy

The International Nuclear Management Academy (INMA) supports universities to establish and deliver master's degree programmes focusing on technology management for the nuclear sector. It provides guidance for master's programmes with a specialized focus on advanced aspects of management and leadership in nuclear technology, science and engineering. INMA targets managers and future managers working in the nuclear energy sector including power and non-power applications.

The IAEA offers support and assistance [add the relevant reference number within this report for the INMA NE Series document] to universities to implement master's programme in NTM and coordinate with the INMA participating universities on the development and maintenance of the requirements for INMA-NTM programmes by organizing meetings and missions.

Any university wishing to have its NTM master's programme recognized as an INMA-NTM programme must first demonstrate that it meets these requirements. Curriculum topics have been identified and defined to help educational institutions design INMA-NTM programmes that are compliant with the requirements.

The INMA curriculum topics are organized into four categories, as described below. It is expected that at least 80% of the curriculum topics are covered by an INMA endorsed NTM programme.

1. External environment that covers political, legal, regulatory, business and societal environmental subjects;

2. Technology that includes the basics of nuclear technology, engineering and applications;

3. Management that addresses the challenges and practices of management in the nuclear and radiological sectors, such as project management, human resource management, and nuclear emergency response;

4. Leadership aligned with strategy, corporate ethics and values.

When the master's programme is recognized as meeting the INMA requirements through a formal IAEA supervised endorsement process, it becomes an IAEA recognized INMA-endorsed NTM programme and the participating university becomes an INMA member. The INMA initial assist mission, final assessment mission and endorsement process are managed by the IAEA. The final assessment determines whether the implementation of a candidate university's NTM programme adequately realizes the INMA-NTM programme requirements. The findings are presented in the INMA final assessment mission report, which also provides observations and suggestions for possible improvements based on the INMA requirements and the collective experience and interpretation of the team members.

Based on the findings and the university's planned follow-up actions to address any needed improvements, the mission team may determine that the programme meets the INMA requirements and recommend to the IAEA that the university be granted INMA member status and that its NTM programme be recognized as an INMA endorsed programme.

This INMA final assessment mission was conducted in response to a request by [name of university, or department and university, or faculty and university] to the IAEA. A team of international experts assembled by the IAEA conducted the assessment of the NTM master's programme offered by the [name of university].

2.2 [name of university]

[Background information and summary of the university]

3. OBJECTIVES

The objectives and deliverables of the assessment mission were to: [list the objectives]

4. ASSESSMENT WORK PROGRAMME

[Timetable of the mission]

5. ASSESSMENT OF THE NUCLEAR TECHNOLOGY MANAGEMENT PROGRAMME AND ITS KEY ELEMENTS

[Summary of the programme including the following and any module specific observations and findings]

5.1 Programme structure

5.2 Key aspects of the courses and how the curriculum topics are taught including the pedagogical approach

5.3 Stakeholder support for the programme

5.4 Overall programme observations and findings

6. RECOMMENDATIONS TO THE IAEA FOR INMA

7. CONCLUSIONS AND RECOMMENDATIONS TO THE UNIVERSITY

8. REFERENCES

    1. The INMA IAEA-NE Series should be referenced.

9. APPENDICES

9.1 Appendix I: Self-assessment tool

9.2 Appendix II: Programme description form

9.3 Appendix III: Course description forms

9.4 Appendix IV: Course delivery and student assessment form

FIG. 8. INMA assessment mission report template.

Appendix VIII

INMA TERMS OF REFERENCE

VIII.1. BACKGROUND

The development of any national nuclear energy programme depends on the successful development of qualified human resources, through sustainable nuclear educational and training programmes supported by government and industry. Among the broad range of specialists needed for the continued safe and economic utilization of all nuclear technologies for peaceful purposes, competent nuclear managers are a vital component of any nuclear workforce.

The International Atomic Energy Agency (IAEA), in resolution GC(58)/RES/13[4], expressed its support to the International Nuclear Management Academy (INMA) initiative in the following terms:

"Encourages the Secretariat to pursue its International Nuclear Management Academy (INMA) initiative, which supports collaborations among nuclear engineering universities around the world to develop a framework for implementing and delivering master's level education programmes in nuclear management, and to facilitate Member State and stakeholder involvement, including financial support for students and course material development".

There are currently few master's degree programmes specializing in management for the nuclear and/or radiological sectors. University master's programmes of business administration, technology management and public administration provide extensive and excellent courses in management, but few of them offer programmes that are specifically applicable to the nuclear and/or radiological sectors. Some existing university departments of nuclear engineering provide courses relating to management, but the number of such departments and courses is limited.

Engineers and scientists at nuclear or radiological facilities have limited opportunities for obtaining formal management education. Likewise, many managers in the nuclear and radiological sectors do not have a nuclear related technical degree and typically have few chances of obtaining such formal nuclear engineering or science education. In many developing countries, particularly those considering or in the process of launching nuclear energy programmes, or those utilizing nuclear non-power applications, there is a lack of both technical and managerial experience in management and leadership roles.

However, effective management and decision making are crucial throughout the nuclear technology life cycle in order to achieve and maintain high levels of safety and performance. Management competencies need to be acquired not only by practical industry focused training courses and on-the-job learning, but also by formal education focused on theory, concepts and academic exercises.

Nuclear and radiological sector professionals who may move into managerial positions in the future, or those who are already managers, whether in developed or in developing countries, are expected to acquire the appropriate skills and knowledge for their positions and this requires both nuclear technology and managerial competencies. To develop these competencies in-house may not be possible as it may be very costly and not as comprehensive as desired. Ideally, managers in the nuclear and radiological sectors should acquire most of the necessary competencies before they move into managerial positions and subsequently continue their professional development.

Both nuclear regulators and nuclear organizations recognize the need and benefits of establishing formal educational programmes in NTM at the master's level to meet this purpose. Such programmes need to be of a high and consistent quality, to be tailored to address the specifics of the nuclear and/or radiological sector, and to be available part-time and through distance learning or short format courses, in order to be accessible to all nuclear professionals. Finally, they need to be available in English to support internationalization of the nuclear workforce and to meet the needs of developing countries.

---

[4] Strengthening the Agency's Activities Related to Nuclear Science, Technology and Applications, Resolution GC(58)/RES/13, IAEA, Vienna (2014).

VIII.2. INMA NUCLEAR TECHNOLOGY MANAGEMENT PROGRAMMES

INMA supports universities to establish and deliver master's degree programmes in nuclear technology management. It is a programme activity of the Nuclear Knowledge Management Section of the Division of Planning, Information and Knowledge Management in the Department of Nuclear Energy of the IAEA. The NTM master's degree programmes significantly benefit Member State nuclear energy activities by strengthening the competencies of technology managers in the nuclear energy sector, including power and non-power applications.

The IAEA supports cooperation among universities in its Member States to develop, implement and deliver INMA-NTM programmes based on the fifty INMA curriculum topics that are defined in this publication (see Section 3.1). Any university authorized by its government to confer a master's degree (including any formally established non-profit organization that is recognized by its government as an institution with a mandate for coordinating a formal collaboration network of such universities for the purpose of joint delivery of a master's degree) can develop a master's degree programme in nuclear technology management that may be endorsed by the IAEA. Endorsed NTM programmes (INMA-NTM programmes) are expected to be substantially (greater than 80%) based on the INMA curriculum topics, as assessed through an INMA assessment missions.

The fifty INMA curriculum topics are organized into four categories, as described below (see Fig. 9 for a full list of the INMA curriculum topics).

(1) External environment: The curriculum topics relating to understanding or managing aspects of the nuclear organization's external environment such as the political, legal, regulatory, business and societal environments in which nuclear managers operate.
(2) Technology: The curriculum topics relating to nuclear technology and engineering, and their applications that are involved directly or indirectly in the management of nuclear facilities for power and non-power applications.
(3) Management: The curriculum topics relating to the challenges and practices of management in the nuclear and radiological sectors with due consideration of safety, security and economics.
(4) Leadership: Requires an understanding of the technology and management of a nuclear facility with due consideration of the external environment in which it operates. The leadership curriculum topics reflect the attributes required for leadership in the nuclear and radiological sectors.

They have been chosen to ensure that INMA-NTM programmes produce an ongoing supply of highly qualified nuclear technology managers for nuclear energy sector employers, including nuclear power plants, waste management facilities, research and development laboratories, regulatory bodies, technical support organizations and nuclear energy related government ministries.

The objective of INMA is to improve the safety, performance and economics of nuclear technologies by promoting and enabling the availability and accessibility of consistent, high quality, educational opportunities for nuclear and radiological sector managers and to improve their management competencies through the following:

(a) Assisting the development, implementation and subsequent assessment of master's level educational programmes in NTM through INMA missions;
(b) Supporting collaborations among universities offering nuclear engineering and science programmes in IAEA Member States to aid the development, implementation and delivery of INMA-NTM programmes;
(c) Integrating employees' nuclear industry experience with formal academic education;
(d) Encouraging the availability and accessibility of high quality INMA-NTM programmes worldwide, through various mechanisms, including but not limited to, collaboration among nuclear universities and resource sharing, e-learning, distance education, part-time programmes, short format courses and innovative use of technology;
(e) Encouraging IAEA Member States to recognize the importance of NTM professionals in achieving and maintaining high levels of safety and performance;

Category 1. External environment
1.1 Energy production, distribution and markets
1.2 International nuclear and radiological organizations
1.3 National nuclear technology policy, planning and politics
1.4 Nuclear standards
1.5 Nuclear and radiological law
1.6 Business law and contract management
1.7 Intellectual property management
1.8 Nuclear and radiological licensing, licensing basis and regulatory processes
1.9 Nuclear security
1.10 Nuclear safeguards
1.11 Transport of nuclear goods and materials

Category 2. Technology
2.1 Nuclear or radiological facility design principles
2.2 Nuclear or radiological facility operational systems
2.3 Nuclear or radiological facility life management
2.4 Nuclear or radiological facility maintenance processes and programmes
2.5 Systems engineering for nuclear or radiological facilities
2.6 Nuclear safety principles and analysis
2.7 Radiological safety and protection
2.8 Nuclear reactor physics and reactivity management
2.9 Nuclear fuel cycle technologies
2.10 Radioactive waste management and disposal
2.11 Nuclear or radiological facility decommissioning
2.12 Environmental protection, monitoring and remediation
2.13 Nuclear research and development and innovation management
2.14 Applications of nuclear science
2.15 Thermohydraulics

Category 3: Management
3.1 Nuclear engineering project management
3.2 Management systems in nuclear or radiological organizations
3.3 Management of employee relations in nuclear or radiological organizations
3.4 Organizational human resource management and development
3.5 Organizational behaviour
3.6 Financial management and cost control in nuclear or radiological organizations
3.7 Information and records management in nuclear or radiological organizations
3.8 Training and human performance management in nuclear or radiological organizations
3.9 Performance monitoring and organization improvement
3.10 Nuclear quality assurance programmes
3.11 Procurement and supplier management in nuclear or radiological organizations
3.12 Nuclear safety management and risk informed decision making
3.13 Nuclear incident management, emergency planning and response
3.14 Operating experience feedback and corrective action processes
3.15 Nuclear security programme management
3.16 Nuclear safety culture
3.17 Nuclear events and lessons learned
3.18 Nuclear knowledge management

Category 4: Leadership
4.1 Strategic leadership
4.2 Ethics and values of a high standard
4.3 Internal communication strategies for leaders in nuclear or radiological organizations
4.4 External communication strategies for leaders in nuclear in nuclear or radiological organizations
4.5 Leading change in nuclear or radiological organizations
4.6 Leadership to support the safety culture

Fig. 9. INMA curriculum topics by category.

The role of the IAEA, in cooperation with INMA members and stakeholders includes the following points:

— Developing and continually reviewing the curriculum topics on which the INMA-NTM programmes are based;
— Organizing missions to assist universities in the development of INMA-NTM programmes and subsequently their compliance assessment;
— Identifying how and when INMA and INMA-NTM programme improvements can be made;
— Hosting, or delegating the hosting of, the annual meeting for INMA members and interested universities (see Section 6)

When a university has its NTM programme assessed to be INMA compliant, the programme may be endorsed by the IAEA as an INMA-NTM programme, and the university may be referred to as an INMA member.

VIII.3. INMA MISSIONS AND ENDORSEMENT PROCESS

INMA missions have been designed to help universities to understand, design and establish INMA-NTM programmes and also to formally assess them with a view to endorsement, leading to the university becoming an INMA member (see Section 4). The experts selected by the IAEA for the mission teams will have experience in the development of a NTM programme at their university or have had significant input into a university's NTM's programme or the establishment of INMA.

An INMA assist mission can only be initiated after the IAEA receives a formal request from the university through the Member State's official channels. As well as providing guidance on programme design, assist missions can also initially include the experiences of other universities that have implemented an INMA-NTM programme or are in the process of doing so.

Following receipt of the mission report compiled on conclusion of an assist mission, the university can identify any possible further actions and resources that will be required for it to develop an INMA-NTM programme. Prior to the subsequent mission that will assess their programme, the university compiles a complete information package consisting of a detailed programme description and background information for submission to the IAEA, (see Section 4.3.1). This mission assesses whether the implementation of their NTM programme is INMA compliant and demonstrates the highest standards of professional conduct, as befitting recognition as an INMA member. The findings and conclusions of the assessment, including any recommendations for possible improvements, as determined by the collective experience of the mission team, are presented in a report coordinated by the IAEA Scientific Secretary.

An action plan, agreed with the university, may be required to address any weaknesses or omissions. If no action plan is required, or only minor issues were found, the assessment report may recommend the university for INMA membership.

If the Director of the Division of Planning, Information and Knowledge Management, in consultation with the Nuclear Knowledge Management Section Head, concurs that the programme is INMA compliant based on the conclusions of the assessment report and any required action plan, the IAEA can accept the recommendation for INMA membership. The university's programme may then be endorsed as an INMA-NTM programme and the IAEA shall confirm this through a formal letter sent to the university with an INMA-NTM programme certificate. Until such letter is received, the university should not assume that it is an INMA member.

For an INMA member to maintain its status and endorsement of its programme, a complete information package must be updated and submitted to the IAEA for review and evaluation every four years. Every eight years, the re-endorsement evaluation must include an assessment mission. All re-endorsements will be signified by a formal letter from the IAEA.

If it is concluded that an INMA member's NTM programme is no longer INMA compliant, the university is not meeting the highest standards of professional conduct as befitting an INMA member, any agreed action plan has not been implemented satisfactorily, or for any other reason the IAEA deems sufficient, the INMA membership and programme endorsement may be revoked.

VIII.4. INMA ANNUAL MEETINGS

The INMA annual meeting is hosted by either the IAEA or one of the INMA members. Attendees of the meeting may include INMA members, interested universities and other stakeholders. INMA members are expected to submit an annual status report (see Section 6) and share their experience on implementing and delivering INMA-NTM programmes with other participants of the meeting. This provides valuable information on the status and development of INMA-NTM programmes including the endorsement process. The IAEA records the progress made by the universities in the implementation of their INMA-NTM programmes as well as any feedback and suggestions for further development based on discussions at the annual meeting.

REFERENCES

[1] Strengthening the Agency's Activities Related to Nuclear Science, Technology and Applications, Resolution GC(58)RES/13, IAEA, Vienna (2014).

[2] INTERNATIONAL ATOMIC ENERGY AGENCY, Status and Trends in Nuclear Education, IAEA Nuclear Energy Series No. NG-T-6.1, IAEA, Vienna (2011).

[3] INTERNATIONAL ATOMIC ENERGY AGENCY, Nuclear Engineering Education: A Competence Based Approach to Curricula Development, IAEA Nuclear Energy Series No. NG-T-6.4, IAEA, Vienna (2014).

[4] INTERNATIONAL ATOMIC ENERGY AGENCY, Selection, Competency Development and Assessment of Nuclear Power Plant Managers, IAEA-TECDOC-1024, IAEA, Vienna (1998).

[5] INTERNATIONAL ATOMIC ENERGY AGENCY, Managing Regulatory Body Competence, IAEA Safety Report Series No. 79, IAEA, Vienna (2013).

[6] INTERNATIONAL ATOMIC ENERGY AGENCY, Training the Staff of the Regulatory Body for Nuclear Facilities: A Competency Framework, IAEA-TECDOC-1254, IAEA, Vienna (2001).

[7] INTERNATIONAL ATOMIC ENERGY AGENCY, Leadership and Management for Safety, IAEA Safety Standards Series No. GSR Part 2, IAEA, Vienna (2016).

[8] INTERNATIONAL ATOMIC ENERGY AGENCY, Milestones in the Development of a National Infrastructure for Nuclear Power, IAEA Nuclear Energy Series No. NG-G-3.1 (Rev. 1), IAEA, Vienna (2015).

[9] INTERNATIONAL ATOMIC ENERGY AGENCY, Governmental, Legal and Regulatory Framework for Safety, IAEA Safety Standards Series No. GSR Part 1 (Rev. 1), IAEA, Vienna (2016).

[10] INTERNATIONAL ATOMIC ENERGY AGENCY, The International Legal Framework for Nuclear Security, IAEA International Law Series No. 4, IAEA, Vienna (2011).

[11] INTERNATIONAL ATOMIC ENERGY AGENCY, Legal Framework for IAEA Safeguards, IAEA, Vienna (2013).

[12] The Convention on the Physical Protection of Nuclear Material, INFCIRC/274/Rev.1, IAEA, Vienna (1980).

[13] INTERNATIONAL ATOMIC ENERGY AGENCY, Nuclear Security Recommendations on Physical Protection of Nuclear Material and Nuclear Facilities (INFCIRC/225/Revision 5), IAEA Nuclear Security Series No.13, IAEA, Vienna (2011).

[14] INTERNATIONAL ATOMIC ENERGY AGENCY, Security in the Transport of Radioactive Material, IAEA Nuclear Security Series No. 9, IAEA, Vienna (2008).

[15] INTERNATIONAL ATOMIC ENERGY AGENCY, Development, Use and Maintenance of the Design Basis Threat, IAEA Nuclear Security Series No. 10, IAEA, Vienna (2009).

[16] INTERNATIONAL ATOMIC ENERGY AGENCY, Nuclear Security Culture, IAEA Nuclear Security Series No. 7, IAEA, Vienna (2008).

[17] INTERNATIONAL ATOMIC ENERGY AGENCY, Security of Nuclear Information, IAEA Nuclear Security Series No. 23-G, IAEA, Vienna (2015).

[18] INTERNATIONAL ATOMIC ENERGY AGENCY, International Safeguards in the Design of Nuclear Reactors, IAEA Nuclear Energy Series No. NP-T-2.9, IAEA, Vienna (2014).

[19] INTERNATIONAL ATOMIC ENERGY AGENCY, International Safeguards in Nuclear Facility Design and Construction, IAEA Nuclear Energy Series No. NP-T-2.8, IAEA, Vienna (2013).

[20] INTERNATIONAL ATOMIC ENERGY AGENCY, Regulations for the Safe Transport of Radioactive Material, IAEA Safety Standards Series No. SSR-6 (Rev. 1), IAEA, Vienna (2018).

[21] INTERNATIONAL ATOMIC ENERGY AGENCY, Safety of Nuclear Power Plants: Design, IAEA Safety Standards Series No. SSR-2/1 (Rev. 1), IAEA, Vienna (2016).

[22] INTERNATIONAL ATOMIC ENERGY AGENCY, Safety of Nuclear Power Plants: Commissioning and Operation, IAEA Safety Standards Series No. SSR-2/2 (Rev. 1), IAEA, Vienna (2016).

[23] INTERNATIONAL ATOMIC ENERGY AGENCY, Safety Assessment for Facilities and Activities, IAEA Safety Standards Series No. GSR Part 4 (Rev. 1), IAEA, Vienna (2016).

[24] INTERNATIONAL ATOMIC ENERGY AGENCY, Establishing the Nuclear Security Infrastructure for a Nuclear Power Programme, IAEA Nuclear Security Series No. 19, IAEA, Vienna (2013).

ABBREVIATIONS

ALARA	As low as reasonably achievable
INMA	International Nuclear Management Academy
IRIDM	Integrated risk informed decision making
OECD/NEA	OECD Nuclear Energy Agency
NTM	nuclear technology management
OECD	Organisation for Economic Co-operation and Development
WANO	World Association of Nuclear Operators

CONTRIBUTORS TO DRAFTING AND REVIEW

Adachi, F.	International Atomic Energy Agency
Ahmadi, S.J.	Nuclear Science and Technology Research Institute, Islamic Republic of Iran
Ariyanto, S.	Centre for Education and Training, Indonesia
Bamford, S.A.	University of Ghana, Ghana
Bastos, J.	International Atomic Energy Agency
Beeley, P.	Khalifa University of Science, Technology and Research, United Arab Emirates
Bhatti, M.A.	Pakistan Atomic Energy Commission, Pakistan
Bischoff, G.	Consultant
Borio, A.	International Atomic Energy Agency
Bruhn, F.	International Atomic Energy Agency
Cagnazzo, M.	University Institute of Advanced Study of Pavia, Italy
Chirlesan, D.	University of Pitesti, Romania
Chmielewski, A.	Institute of Nuclear Chemistry and Technology, Poland
Clark, R.	Consultant
Coeck, M.	Belgian Nuclear Research Centre, Belgium
Cortes, G.P.	Universitat Politècnica de Catalunya, Spain
de Grosbois, J.	International Atomic Energy Agency
Day, J.	Consultant
Demachi, K.	University of Tokyo, Japan
Dies, J.	Technical University of Catalonia, Spain
Drury, D.	International Atomic Energy Agency
Froats, J.	University of Ontario Institute of Technology, Canada
Gao, P.Z.	Harbin Engineering University, China
Gauthier, J-C.	AREVA, France
Geraskin, N.	National Research Nuclear University, Moscow Engineering Physics Institute, Russian Federation
Gevorgyan, A.	National Polytechnic University of Armenia, Armenia
Glöckler, O.	International Atomic Energy Agency
Gulley, N	International Atomic Energy Agency
Gysel, T.	Consultant, Switzerland
Hanamitsu, K.	International Atomic Energy Agency

Hirose, H.	International Atomic Energy Agency
Jaghoub, M.	University of Jordan, Jordan
Jiang, F.	International Atomic Energy Agency
Jorant, G.	Consultant
Karezin, V.	State Corporation Rosatom, Russian Federation
Khaidzir, H.	Universiti Teknologi Malaysia, Malaysia
Ki In Han	KEPCO International Nuclear Graduate School, Republic of Korea
Kosilov, A.	National Research Nuclear University Moscow Engineering Physics Institute, Russian Federation
Kurwitz, C.	Texas A&M University, United States of America
Kusumi, R.	Consultant
Larkin, J.	University of the Witwatersrand, South Africa
Lee, T.J.	Korea Atomic Energy Research Institute, Republic of Korea
Leslie, I.	World Nuclear University
Liu, L.	International Atomic Energy Agency
Myung Jae Song	Seoul National University, Republic of Korea
Nagibina, E.	State Corporation Rosatom, Russian Federation
Nilsuwankosit, S.	Chulalongkom University, Thailand
Oda, T.	Seoul National University, Republic of Korea
Ravetto, P.	Politecnico di Torino, Italy
Ricotti, M.	Politecnico di Milano, Italy
Roberts, J.W.	The University of Manchester, United Kingdom
Shi, L.	Tsinghua University, China
Stafford, I.	World Nuclear Association
Sun, Y.K.	University of South China, China
Sun, Y.L.	Tsinghua University, China
Tarren, P.	International Atomic Energy Agency
Thi Thu Trang Pham	Vietnam Atomic Energy Institute, Viet Nam
Thomauske, B.	RWTH Aachen University, Germany
Tshivhase, M.	North-West University, South Africa
Uesaka, M.	University of Tokyo, Japan
Van Wyk, L.	North-West University, South Africa

Volkov, Y.	National Research Nuclear University Moscow Engineering Physics Institute, Russian Federation
Wichers, H.	North-West University, South Africa
Wieland, P.	World Nuclear University
Yoo, S.J.	Korea Nuclear International Cooperation Foundation, Republic of Korea
Zhao, L.,	Tsinghua University, China
Zimmermann, D.R.	International Atomic Energy Agency

Technical Meetings

Trieste, Italy, 28–31 July 2015
Manchester, United Kingdom, 3–6 May 2016

Consultants Meetings and Missions

Vienna, Austria, 25–27 November 2013; 28–30 April 2014
Tokyo, Japan, 9–13 June 2014
Manchester, United Kingdom, 29–31 July 2014
Vienna, Austria, 24–27 November 2014
Texas, United States of America, 6–9 October 2014
Moscow, Russian Federation, 28–31 October 2014
Johannesburg, South Africa, 23–24 February 2015
Potchefstroom, South Africa, 25–26 February 2015
Beijing, China, 31 March–3 April 2015
Manchester, United Kingdom, 30 June–3 July 2015
Trieste, Italy, 27–31 July 2015
Vienna, Austria, 16–19 February 2016
Manchester, United Kingdom, 3–6 May 2016
Oshawa, Canada, 10–14 May 2016

Structure of the IAEA Nuclear Energy Series*

Nuclear Energy Basic Principles
NE-BP

Nuclear Energy General Objectives
NG-O

1. Management Systems
NG-G-1.#
NG-T-1.#

2. Human Resources
NG-G-2.#
NG-T-2.#

3. Nuclear Infrastructure and Planning
NG-G-3.#
NG-T-3.#

4. Economics and Energy System Analysis
NG-G-4.#
NG-T-4.#

5. Stakeholder Involvement
NG-G-5.#
NG-T-5.#

6. Knowledge Management
NG-G-6.#
NG-T-6.#

Nuclear Reactor Objectives**
NR-O

1. Technology Development
NR-G-1.#
NR-T-1.#

2. Design, Construction and Commissioning of Nuclear Power Plants
NR-G-2.#
NR-T-2.#

3. Operation of Nuclear Power Plants
NR-G-3.#
NR-T-3.#

4. Non Electrical Applications
NR-G-4.#
NR-T-4.#

5. Research Reactors
NR-G-5.#
NR-T-5.#

Nuclear Fuel Cycle Objectives
NF-O

1. Exploration and Production of Raw Materials for Nuclear Energy
NF-G-1.#
NF-T-1.#

2. Fuel Engineering and Performance
NF-G-2.#
NF-T-2.#

3. Spent Fuel Management
NF-G-3.#
NF-T-3.#

4. Fuel Cycle Options
NF-G-4.#
NF-T-4.#

5. Nuclear Fuel Cycle Facilities
NF-G-5.#
NF-T-5.#

Radioactive Waste Management and Decommissioning Objectives
NW-O

1. Radioactive Waste Management
NW-G-1.#
NW-T-1.#

2. Decommissioning of Nuclear Facilities
NW-G-2.#
NW-T-2.#

3. Environmental Remediation
NW-G-3.#
NW-T-3.#

(*) as of 1 January 2020
(**) Formerly 'Nuclear Power' (NP)

Key
BP: Basic Principles
O: Objectives
G: Guides and Methodologies
T: Technical Reports
Nos 1–6: Topic designations
#: Guide or Report number

Examples
NG-G-3.1: Nuclear Energy General (**NG**), Guides and Methodologies (**G**), Nuclear Infrastructure and Planning (topic **3**), **#1**
NR-T-5.4: Nuclear Reactors (**NR**)*, Technical Report (**T**), Research Reactors (topic **5**), **#4**
NF-T-3.6: Nuclear Fuel (**NF**), Technical Report (**T**), Spent Fuel Management (topic **3**), **#6**
NW-G-1.1: Radioactive Waste Management and Decommissioning (**NW**), Guides and Methodologies (**G**), Radioactive Waste Management (topic **1**) **#1**

**IAEA**
International Atomic Energy Agency

ORDERING LOCALLY

IAEA priced publications may be purchased from the sources listed below or from major local booksellers.

Orders for unpriced publications should be made directly to the IAEA. The contact details are given at the end of this list.

NORTH AMERICA

Bernan / Rowman & Littlefield

15250 NBN Way, Blue Ridge Summit, PA 17214, USA
Telephone: +1 800 462 6420 • Fax: +1 800 338 4550

Email: orders@rowman.com • Web site: www.rowman.com/bernan

REST OF WORLD

Please contact your preferred local supplier, or our lead distributor:

Eurospan Group

Gray's Inn House
127 Clerkenwell Road
London EC1R 5DB
United Kingdom

Trade orders and enquiries:

Telephone: +44 (0)176 760 4972 • Fax: +44 (0)176 760 1640
Email: eurospan@turpin-distribution.com

Individual orders:

www.eurospanbookstore.com/iaea

For further information:

Telephone: +44 (0)207 240 0856 • Fax: +44 (0)207 379 0609
Email: info@eurospangroup.com • Web site: www.eurospangroup.com

Orders for both priced and unpriced publications may be addressed directly to:

Marketing and Sales Unit
International Atomic Energy Agency
Vienna International Centre, PO Box 100, 1400 Vienna, Austria
Telephone: +43 1 2600 22529 or 22530 • Fax: +43 1 26007 22529
Email: sales.publications@iaea.org • Web site: www.iaea.org/publications